01209 616182

LEARNING.
•••••••••••••services

Cornwall College Camborne
Learning Centre - HE

This resource is to be returned on or before the last date
stamped below. To renew items please contact the Centre

12 issues '08

Three Week Loan

Practical Statistic:
Experimental Bic

Second Edition

Practical Statistics for Experimental Biologists

Second Edition

ALASTAIR C. WARDLAW

Emeritus Professor of Microbiology
University of Glasgow, UK

MINITAB is a registered trademark of MINITAB Ltd.

JOHN WILEY & SONS, LTD

Chichester · New York · Weinheim · Brisbane · Singapore · Toronto

Other Wiley Editorial Offices

John Wiley & Sons Inc., 605 Third Avenue,
New York, NY 10158-0012, USA

WILEY-VCH Verlag GmbH, Pappelallee 3,
D-69469 Weinheim, Germany

Jacaranda Wiley Ltd, 33 Park Road, Milton,
Queensland 4064, Australia

John Wiley & Sons (Asia) Pte Ltd, 2 Clementi Loop #02-01,
Jin Xing Distripark, Singapore 129809

John Wiley & Sons (Canada) Ltd, 22 Worcester Road,
Rexdale, Ontario M9W 1L1, Canada

Library of Congress Cataloging-in-Publication Data
Wardlaw, A. C.
    Practical statistics for experimental biologists / Alastair C.
Wardlaw.—2nd ed.
      p.   cm.
    Includes bibliographical references.
    ISBN 0-471-98821-9 (alk. paper).—ISBN 0-471-98822-7 (pbk. : alk. paper)
      1. Biometry.   2. Biology, Experimental—Statistical methods.
    I. Title.
    QH323.5.W37   1999
    570′.7′27—dc21
                                                              99–35926
                                                                  CIP
British Library Cataloguing in Publication Data

A catalogue record for this book is available from the British Library

ISBN 0-471-98821-9 (PPC)
ISBN 0-471-98822-7 (Pbk)

Typeset in 10/12pt Times by Footnote Graphics, Warminster, Wilts.
Printed and bound in Great Britain by Antony Rowe Ltd, Chippenham, Wiltshire..

This book is printed on acid-free paper responsibly manufactured from sustainable forestry,
in which at least two trees are planted for each one used for paper production.

Contents

Preface

This preface is short – with the thought that it might be read!

The First Edition (1985) of *Practical Statistics* expected the reader to put data into algebraic equations with a pocket calculator and look-up statistical tables – the traditional approach of the time. Nowadays, computer-based statistics packages do not display equations or statistical tables and they mostly seem to assume that the user is expert in statistical terminology. Personally, I have run into numerous difficulties with the 'jargon blizzard'. And I have frequently felt totally inadequate in not understanding something that the package manuals imply is 'easy'. I know other people who have struggled similarly. It seems there is quite a gap between finding out about statistical methods and ideas from a traditional textbook, and putting them into practice on the screen.

This Second Edition aims to bridge this gap. It assumes the reader wants to use statistics in a responsible way, but has little interest in processing data with formulae and a pocket calculator. It is therefore for those who want to go onto the computer screen straightaway and become familiar with the statistical methods that can be used there – and the basic ideas behind them. The text is intended to be worked through as a course, and thereafter used as a reference manual.

Only after many enquiries did I decide to settle on MINITAB* as the particular statistics package to be used throughout the revised text. There are several other good statistics packages that I could have chosen, but MINITAB has become so widely available and accepted as a standard that anyone with access to a PC or Apple computer should also have access to MINITAB. At least that is my opinion. The pocket calculator of yesterday has become the MINITAB of today and tomorrow.

I want to acknowledge help from statistician colleagues at the University of Glasgow, particularly Professor Adrian Bowman, Dr James Currall and Dr Jim Kay. However, any faults are my doing and responsibility, and not theirs. I am also much indebted to MINITAB Ltd for permission to copy from their software and for my enrolment in their Authors' Assistance Program. Other permissions to use copyright materials are acknowledged at appropriate places in the text. Finally, I thank the publishers for their guidance and patience, and hope that you, the reader, get some benefit from what follows.

A. C. W.
Glasgow, Scotland, 1999

*MINITAB Ltd, 3 Mercia Business Village, Torwood Close, Westwood Business Park, Coventry CV3 5QQ, United Kingdom. Tel.: (+44) 01203 695730; Fax.: (+44) 01203 695731; email: ltd@minitab.co.uk; World-wide web: http://www.minitab.com

1

A Simple Experiment in Pipetting

The notion that experiments and other research investigations can be conducted statistically or non-statistically, at the will of the investigator, is firmly held by many: it is usually entirely false.

D. J. Finney

1.1 COLLECTING DATA

This book starts by asking the reader to go into the laboratory and do a quick 'wet' experiment to check the accuracy of that basic tool, the automatic pipette. However, there is a provided data set for those who prefer to 'stay dry'. One way or the other, the main purpose is to have some results that can readily be visualized and then used to introduce the MINITAB statistics package. A byproduct for those who do the actual laboratory work is information on the accuracy of the pipettes in local use. It is an experiment that every biology and biomedical student or researcher should try at least once since it provides a useful overview of numerous statistical procedures and ideas.

More importantly, the weight data from this experiment are representative of the many different types of *measurements* that are generated in the laboratory. Other examples of measurement data are lengths, areas, volumes, times, optical absorbancies, concentrations of metabolites, physiological measurements – in fact, most types of biomedical data except *counts*, which require a different treatment. So the statistical insights developed in this pipetting example should be seen as having wide application. To base the exercise on pipettes is just to make it readily accessible to the experimental bioscientist.

The experiment also illustrates a general approach in this book, which is to take a single set of data and use it to illustrate a wide range of statistical procedures. The alternative and more common practice in textbooks is to use different data sets for each statistical test or theme. It may be a matter of taste, but for me it seems better to change only one variable (the statistical procedure) at a time and keep the data set the same for as long as possible. Then the whole toolkit of statistical methods and ideas can be transferred as a unit to whatever other types of measurement data require analysis.

1.2 STANDARDIZING PIPETTES

Although the pipette is a device for measuring volume, a convenient way to check its accuracy and consistency is by weighing the delivery of a liquid of known specific gravity, such as water at 20°C. An electronic balance of the type common in laboratories can weigh to 0.1 mg, which is 1/1000 of the weight of 100 μl of water, the volume to be used here. The accuracy of the weighing is therefore likely to be much higher than that of the pipette itself, even if the pipette has been properly calibrated and then used with care and consistency. Certainly the pipettes to be found in student laboratories may be woefully inaccurate, as is likely to be revealed by what follows!

1) Take three automatic pipettes (Eppendorf, Finn, Gilson, etc.) of the same size (or of different capacities or types if you like), and adjust each to deliver 100 μl; you can also use glass pipettes or calibrated loops if you prefer and volumes other than 100 μl;
2) Put temporary paper labels 1, 2, or 3 on each pipette so as to be able to distinguish them;
3) Place a small glass or plastic bottle on the balance pan and adjust the scale to zero, i.e. 'tare' the bottle;

4) As carefully as possible deliver 100 µl water and record the weight to the nearest 0.1 mg;

5) Re-zero the balance and continue to make deliveries so as to collect a total of three from each pipette. For extra consistency, it would be best to use a new pipette tip for each delivery.

Assuming that the pipettes are all perfectly accurate and used correctly, and that the water is air-free and at 20°C then, according to standard tables, all the deliveries should weigh 99.8 mg. If all the nine weighings are exactly 99.8 mg, then congratulations! It has never before happened in my experience. So to continue with this chapter, set aside your own data and use instead the figures provided.

1.3 PURPOSE AND DESIGN

1.3.1 Purpose

The purpose of the experiment is to provide a set of results for statistical analysis by MINITAB and to answer the following questions:

1) What is the average weight of the nine deliveries, taken as a single group?
2) What is the average weight of the three deliveries from each pipette?
3) How much scatter, or variation, is there in the nine deliveries taken together?
4) How much scatter, or variation, is there in the three deliveries from each pipette?
5) How *accurate* are the nine deliveries? This is different from Question 3. See FN 2.4 at the end of Chapter 2 for the follow-up on this.
6) Are there *significant differences* in the mean volumes delivered by the three pipettes? Discussion of this is deferred until Chapter 3.
7) How should the experiment be done so as to avoid confusing possible volumetric differences between the pipettes with a potential training or fatigue effect of the operator? For example, using the third pipette more (or less) carefully than the first?

These questions are similar to those that arise frequently in laboratory investigations that yield data consisting of measurements. Therefore the experience gained here will have wide application. First we have to consider how the weight readings should be taken.

1.3.2 Design A (Unsatisfactory)

Set the balance to zero with the empty weighing bottle; then make three deliveries from Pipette No. 1, followed by three from Pipette No. 2, finishing up with three from Pipette No. 3, with weighing and re-zeroing of the balance after each delivery.

This will give a nice tidy array of triplicated results opposite each pipette number. But is it necessarily the best method? The main drawback is that such a design is not the best for answering Question 7, i.e. determining whether differences in average volume delivered by different pipettes are *real*, or might have arisen from the experimenter varying in care or expertise while progressing through the nine deliveries. It would be like taking a person into a shooting alley to fire three shots from each of three rifles. If the shooting was better with the third rifle than with the first, how would you know whether this was due to the last rifle itself being more reliable, or because the shooting improved progressively with successive shots? So you would not be able to tell whether there was a training effect or a rifle effect.

Thus there are potentially two *independent possible sources of variation* – an important and basic statistical idea:

1) Possible variation due to intrinsic physical differences between individual pipettes or rifles; and
2) Possible variation originating in the wobbly hand or imperfect eye of the human operator. Therefore to assess the importance of these two separate sources of variation, the experiment must be done so that the two sources are not confused, or *confounded* (to use the statistician's term). In other words, the two influences – if we think they may affect the results – should not be mixed up in such a way that they can not be separated afterwards.

1.3.3 Design B (Random and better)

This takes account of the above comments and is the method to be used. The essential feature is that the nine deliveries will be made in a *pre-determined random sequence*, such as Pipette No. 2, then No. 3, then No. 3 again, then maybe No. 1, No. 2, No. 1, and so on. How randomness is achieved is described below. It is not just a matter of arranging the sequence to look random. Also note that randomization is not a cure for poor technique. It does not eliminate the possibility of bias; it just helps to reduce it, and is a reasonable routine to be included when designing an experiment.

There are various types of random design. The one to be used here is the *completely random design* where all nine weighings are collected in a pre-determined random sequence. An alternative would be a *randomized-block design* in which one delivery would be made from each pipette in a random sequence to give Block No. 1. The randomization would then be repeated and a second set of three deliveries, one from each pipette, would be collected to give Block No. 2. Then there would be a third round of randomization and pipetting to give Block No. 3. Such designs are best if there is a genuine possibility that the observations might change systematically as the successive observations were obtained. For example, if it was strongly suspected that there was a training or fatigue effect of the operator, it might be better to use a randomized block than a fully randomized design. Then each block would provide a snapshot of operator status as the experiment progressed. Any systematic drift due to training or fatigue could then be separated from possible differences in the pipettes themselves.

At this stage some readers may feel that an unnecessarily elaborate ritual is being suggested for a very straightforward experiment. But the intention is to emphasize the crucial point that if you intend to use statistical methods to analyse the results of a quantitative experiment, then these methods must be built in *before* the experiment is done.

It is an unfortunate fact that some researchers use statistical analysis for data that have been gathered without randomization. This causes statisticians to sigh deeply and wearily! It is one of the reasons why statistics as an intellectual discipline gets a bad reputation: 'You can prove anything with statistics'; 'there are three kinds of lies – lies, damned lies and statistics', etc. The reality is that any tool or device used incorrectly (such as not taking proper precautions) can give bizarre results, and statistical methods are no exception.

So to minimize bias and possible confounding, the nine pipette deliveries will be collected in a *pre-determined random sequence*. As we shall see elsewhere in this book, the experimental scientist has to develop the knack of finding the middle ground between achieving 'full statistical virtue', and doing poorly designed experiments that yield unanalysable data. The issue can not be dismissed with 'a well designed experiment does not need statistics'. If a quantitative experiment is indeed well designed it should include statistical principles such as replication of observations and the minimizing of confounding factors.

1.4 OPENING MINITAB

Before starting the pipetting, there is a necessary diversion into MINITAB and how to get random numbers. The reader is assumed to have a computer, or ready access to one, and that 'in this day and age' the machine has MINITAB installed as a standard feature. Thus the reader is also assumed to have reasonable familiarity with pull-down windows, the scrolling bars, how to respond to dialog boxes, use a mouse, and to open, save and print files. So from here on in this book it is assumed that your computer has MINITAB installed, preferably the recently available (1998) MINITAB Release12, although other late versions should be suitable.

1.4.1 Why MINITAB?

The MINITAB Statistics package has several advantages:

1) Capability of handling very large sets of data, like thousands of observations if need be;
2) A wide range of procedures for recording, displaying and statistically analysing data;
3) High-speed processing, like doing in one second a calculation that might take half an hour with statistical formulae, a pocket calculator, pencil and paper;
4) Said to be the most widely used statistics package (A'Brook & Weyers, 1996);
5) Compatibility with *Excel* and other data-analysis packages.

The main initial disadvantage of MINITAB is its sheer sophistication and complexity, so that guidance is needed to find the right tools for a particular task. The documentation that comes with MINITAB gives the impression that everything is 'easy'. This is like the captain of a jet airliner telling you how easy it is to use all the switches and instruments in the cockpit. It may be easy, but only when you know which ones to use when, and what they do! This book does not attempt to explore all the capabilities of MINITAB, but only to present those features that have been most useful to me as a biomedical experimental scientist. It is assumed here that the reader not only has access to MINITAB Release 12, but also to the two MINITAB *User's Guides* (MINITAB, 1997a, b).

1.4.2 Anatomy of MINITAB

The visible components of MINITAB are:

1) The *Worksheet window* which consists of the array of rows and columns where data are entered and manipulated. This has 100 000 'cells', so an enormous data table can be accommodated.

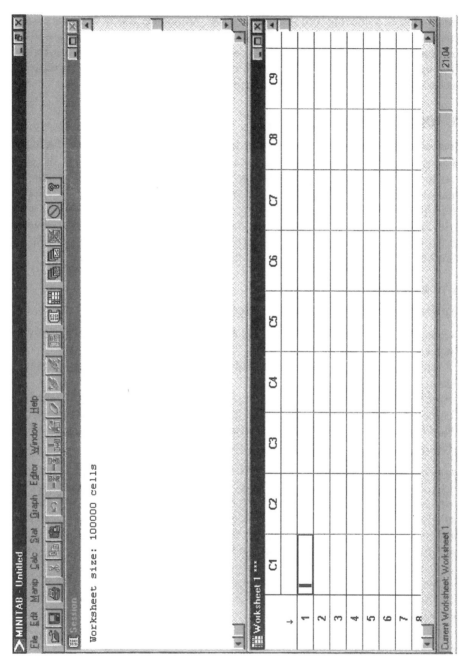

Figure 1.1   MINITAB screen with *Session Window* above and *Worksheet Window* below

2) The *Session window* where the numerical results of some statistical calculations are displayed (i.e. those where the results do not appear on the worksheet). Figure 1.1 shows a MINITAB screen as it normally opens, with the Session window above and the Worksheet window below, both sub-windows being equipped with scroll bars and buttons (top right of each) for reducing, expanding or closing the displays. For example, by pressing the little square in the top right of the Worksheet window, it can be made to occupy the whole screen. To recover the Session window, pull down the *Window* menu at the top of the screen and select *Session Window* for opening. Note that only a very small part of the worksheet can be accommodated on the screen at any one time, so that the scroll bars have to be used for large data sets.

3) The *Graphics windows,* where graphs are presented;

4) The *Toolbar* (on the screen, above the top of Figure 1.1), which carries the pull-down menus for data manipulation and analysis. If you have your MINITAB open, you should see the Toolbar with the following array of labels:

    File   Edit   Manip   Calc   Stat   Graph   Editor   Window   Help

    Try pulling each of these down, and explore their right-sideways extensions, to get an idea of the vast number of possibilities that MINITAB offers.

1.4.3   Storage of data and results

This is done in a filing folder called a *MINITAB project.* The idea is to have a separate MINITAB project for each experiment, or group of similar experiments. Therefore the first step in the pipetting experiment, assuming that MINITAB is installed on the computer, is to open MINITAB itself. This will automatically present a new MINITAB Worksheet and Session window within a new MINITAB project. When the *Save* command is activated in the *File* menu, fill in an appropriate title like *Chap1Pip Exp* and the work will be saved in a MINITAB project with 'mpj' as the file extension. The act of saving a MINITAB project saves everything that is on the worksheet, in the Session window and also any graphs that have been generated.

1.4.4   Statistical formulae and tables

A major difference between this book and the First Edition (and with many recent statistics texts) is the displacement of statistical formulae. However, some will be found in Further Notes sections at the end of the chapters. This has been done because MINITAB applies the appropriate formulae for the actions requested, but without first displaying them. It also provides the relevant sections of statistical tables for interpreting the end results. This book is therefore not intended for mathematically minded readers who may prefer the elegance of an algebraic equation to a series of MINITAB commands. Therefore some formulae *are* given, but generally not as a prerequisite for the MINITAB. Much emphasis, however, is given to the preliminary inspection of experimental results by pictorial displays. Very often these give a very good approximation of the conclusions that will eventually emerge from the full statistical analysis.

The reader is encouraged in this way to make early use of the power and speed of MINITAB and without necessarily serving an apprenticeship in statistical formulae, pocket calculators and statistical tables. Some statisticians do not approve of this approach but I believe it has sufficient merits to outweigh the drawbacks. Readers will decide for themselves.

1.5   ARRANGING A RANDOM SEQUENCE

For collecting the nine pipette deliveries, an array of the numbers 1, 2, and 3 are needed in a random sequence, from which can be chosen three of the number 1, three of the number 2 and three of the number 3. But although only nine random numbers are needed altogether, MINITAB will be asked to produce 20, so as to have some spares. This allows for uneven repetition of the same digit that is an inescapable feature of randomness. For example, the randomness command might deliver ten successive digits all of the number 1, but the chance of this is very small (1 in 3^{10}, or about 1 in 60 000).

The 20 random numbers 1, 2, and 3 are to be delivered into Column 1 (C1) of the Worksheet, so the head of that column should be labelled with *RN* for random numbers.

1.5.1   Getting random numbers

As illustrated in Figure 1.2, go to **Calc > Random Data > Integer** and, when the dialog box *Integer Distribution* appears, use the Tab key, Mouse and $\boxed{Select}$ button to compose Figure 1.3:

1) Insert $\boxed{20}$ between $\boxed{Generate}$ and $\boxed{rows\ of\ data}$; this tells MINITAB that 20 random digits are wanted;
2) Choose $\boxed{C1}$ or $\boxed{\text{'RN'}}$ (the quotes are essential if you use RN) as the *Store in columns(s):* entry for the random numbers; the selection can be done in three ways; highlighting and double clicking, highlighting and clicking on the $\boxed{Select}$ button or direct typing ;
3) Insert $\boxed{1}$ opposite $\boxed{Minimum}$ value;
4) Insert $\boxed{3}$ opposite $\boxed{Maximum}$ value. These last two steps tell MINITAB that the digits 1, 2 and 3 are wanted.
5) Finally click on $\boxed{OK}$.
6) Check that the *RN* column now contains a set of 20 digits, 1, 2 and 3 in a disorderly sequence (Figure 1.4).

1.5.2   Nature of randomness

Statisticians are very fussy about randomness since it underpins the logic of statistical analysis. If the *RN* column is regarded as having initially 20 empty spaces, then the idea of randomness is that each space:

1) is independent and entirely uninfluenced by the contents of the others;
2) has an equal probability of receiving a 1, a 2 or a 3.

This means that in a long run of, say, 1000 random digits you would expect to have very close to one-third of each of 1, 2, and 3. But in a short run of, say, six random

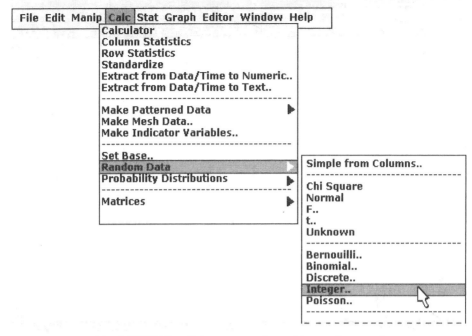

Figure 1.2   MINITAB pulldown menus for getting random numbers

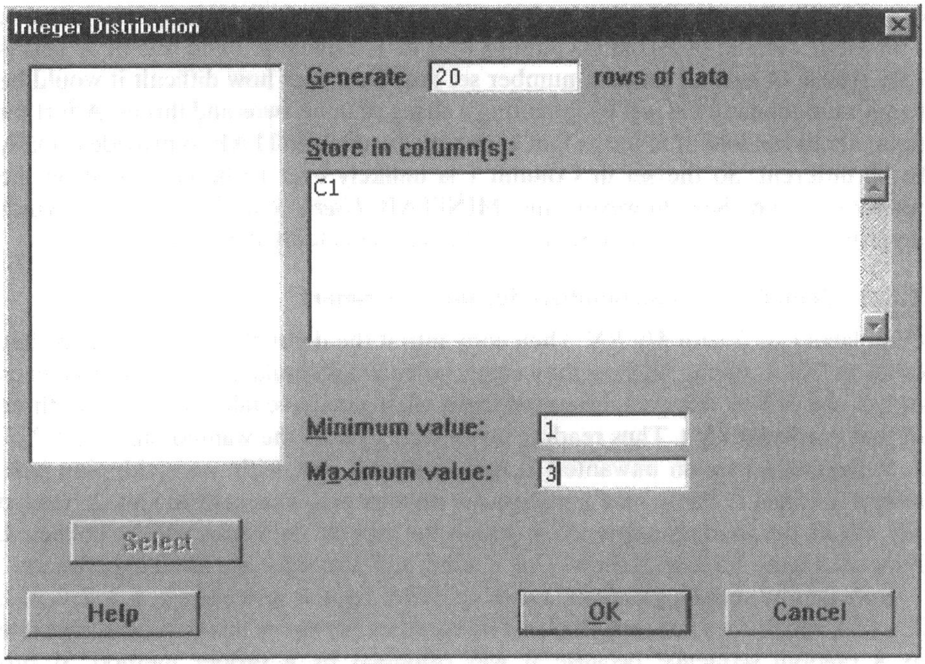

Figure 1.3   MINITAB dialog box for generating 20 of the random digits 1, 2, and 3

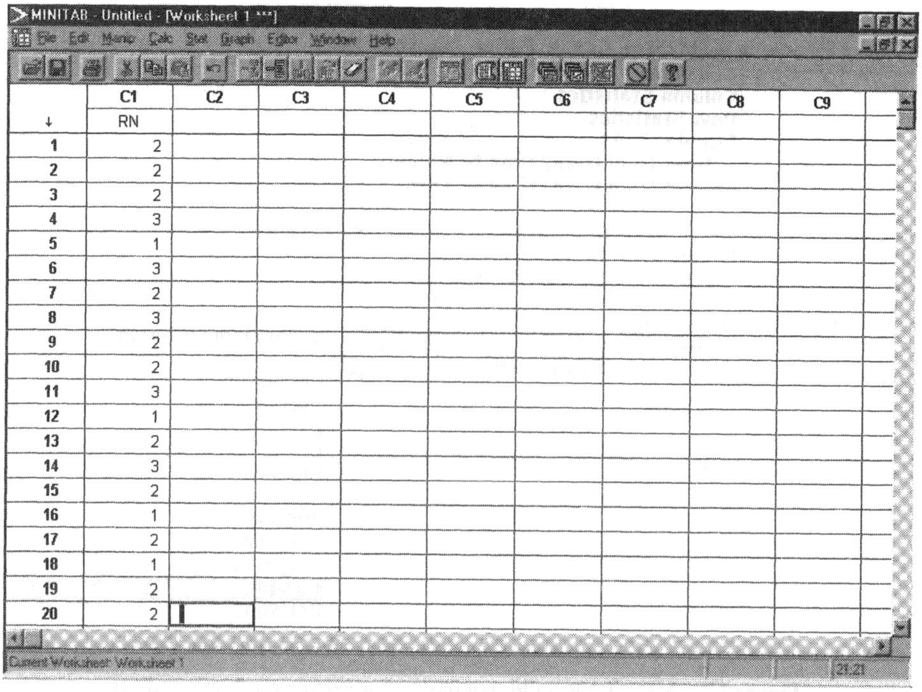

Figure 1.4 Worksheet with a set of 20 random numbers 1, 2, and 3 generated by MINITAB

numbers you are not very likely to get exactly two of 1, two of 2 and two of 3. For example, the first six digits in the *RN* column in Figure 1. 4 are 2, 2, 2, 3, 1, and 3. The whole column of 20 digits contains four of 1, eleven of 2 and five of 3. This is very typical of a small random-number set and illustrates how difficult it would be to get random numbers just by inventing a string of ones, twos and threes. A further point about random numbers is that each set you ask MINITAB to provide is likely to be different. So the set in Column 1 is unlikely ever to be generated on the reader's screen. See, however, the MINITAB *User's Guide No.1* §9.3, which describes how to get the same random numbers repeatedly if wanted.

1.5.3 Choosing random numbers for the experiment

First, label Col. 2 with *My RN*. Then copy into it the digits that are needed as they occur in Col. 1, taking them as they come, so as to accumulate a total of three each of 1, 2, and 3. You skip past unwanted digits when you have filled the quota of three of that particular digit. Thus reading down Col. 1 yields the wanted digits 2, 2, 2, 3, 1, 3, then skip past an unwanted 2; then continue and, with more skipping past, collect 3, 1, and 1. These nine numbers are now entered vertically in Col. 2, headed *My RN* as the random sequence in which the pipette deliveries will be collected. Now see Figure 1.5, which shows Col. 2 filled, and also the completed experiment.

Note that by chance, you could finish up with a regular series like 1, 1, 1, 2, 2, 2, 3, 3, and 3, which does not look the least bit random. However this is quite acceptable as a random sequence because it was obtained by a proper method. If this happened in 'real-life' laboratory work, you would still be justified in accepting it.

↓	C1 RN	C2 My RN	C3	C4 Wts	C5 Pip No	C6 Wt mg	C7	C8 Accurate	C9
1	2	2		97.2	1	100.3		1	
2	2	2		98.5	1	99.7		1	
3	2	2		97.2	1	100.7		1	
4	3	3		97.6	2	97.2		0	
5	1	1		100.3	2	98.5		0	
6	3	3		99.8	2	97.2		0	
7	2	3		99.5	3	97.6		0	
8	3	1		99.7	3	99.8		1	
9	2	1		100.7	3	99.5		1	
10	2								
11	3								
12	1								
13	2								
14	3								
15	2								
16	1								
17	2								
18	1								
19	2								
20	2								

Figure 1.5 MINITAB worksheet with random numbers, weights and sorted weights. The filling of Col. 6 '*Accurate*' is described in FN 2.4

This is because most experiments will be repeated several times before reaching final conclusions, and you would be most unlikely in a run of, say, five repeat experiments to have any repetitions of a particular set of random numbers.

For additional information about random numbers, see the Further Notes (FN 1.1) at the end of this chapter.

1.6 COLLECTING AND SORTING THE OBSERVATIONS

With the first two columns of the worksheet filled, a printout should be taken into the laboratory to record the weights of the nine deliveries of 100 μl water from each of the three pipettes set to deliver that volume. The weights obtained in the sample experiment are recorded in mg in Col. 4 of Figure 1.5, labelled with *Wts* at the top. These weights should then be entered into the worksheet in the computer.

1.6.1 Unscrambling the randomization

Having collected the experimental observations in a pre-determined random sequence, the next step is to unscramble the randomization and present a systematic tabulation. There is no problem of shortage of space on a MINITAB worksheet, therefore Col. 3 can be left as an empty, spacer, column for cosmetic purposes, and the retabulation done in Cols 5 and 6. Since MINITAB does not allow two columns on a worksheet to have the same label, the designations *Pip No* (Pipette Number) for Col. 5 and *Wt mg* for Col. 6 are used.

In Col. 5 the entries 1, 1, 1, 2, 2, 2, 3, 3, 3 are needed for the pipette numbers, so that the corresponding weights can be entered opposite them in Col. 6. This can be done either by typing the numbers directly, or automatically with MINITAB's *Make Patterned Data* command. With only nine entries it is quickest to do it manually, but with a larger data table it would make sense to use the MINITAB command. So to introduce the method, MINITAB will be used.

1.6.2 Making patterned data

Go to **Calc > Make Patterned Data > Simple Set of Numbers**, and when the dialog box (Figure 1.6) appears, fill the spaces as:

1) *Store patterned data in:* ⌐'Pip No'⌐;
2) *From first value:* ⌐1⌐;
3) *To last value:* ⌐3⌐;
4) *In steps of:* ⌐1⌐;
5) *List each value:* ⌐3⌐ *times;*
6) *List the whole sequence:* ⌐1⌐ *times;*
7) ⌐*OK*⌐.

This should produce the three each of 1s, 2s, and 3s as shown in Col. 5. Now for the unscrambling of the randomization, again this could be done manually but MINITAB will do it automatically.

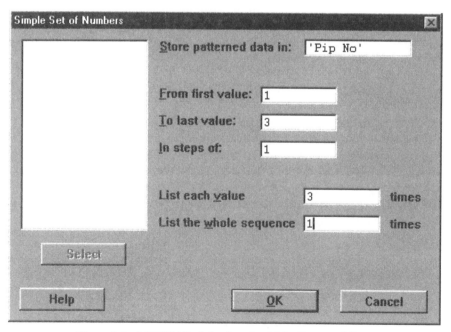

Figure 1.6   MINITAB dialog box for *Make Patterned Data*

Go to: **Manip > Sort**, and when the dialog box (Figure 1.7) appears, fill the spaces as follows:

---

1) *Sort column(s):* Wts (Note, no quotes);

2) *Store sorted column(s) in:* 'Wt mg' (Note, quotes needed);

3) *Sort by column:* 'My RN' *Descending* (leave the adjacent small box unticked, so that the sorting will by default be *ascending*, from 1 to 3).

4) OK.

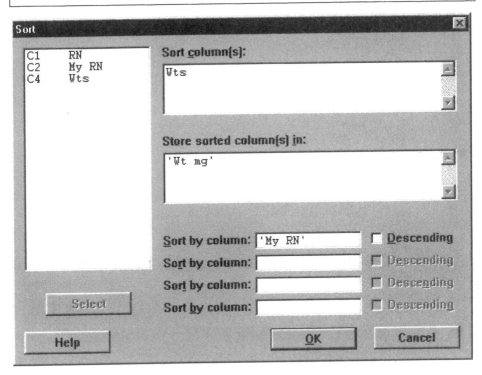

Figure 1.7   MINITAB dialog box for *Sort*

This should give the systematically-tabulated weights of the volumes delivered by each pipette in Col. 6 of Figure 1.5. The above few paragraphs may seem to be overly elaborate in what they expect the experimenter to do for a simple end-result, but this is intended mainly as an introductory example for gaining experience with MINITAB.

1.7   PICTORIAL DISPLAY

1.7.1   Point plot

Having obtained the data in systematic form, it is good practice first to get a pictorial display in the form of a *point plot*. MINITAB offers the *Dotplot* option in its *Graph* pull-down window. However, there is a more useful plot of points, with more flexibility for editing, if the *Plot* option is taken.

Therefore go to **Graph > Plot**, and when the dialog box (Figure 1.8, top) appears:

1) At *Graph variables* and opposite *Graph 1*, enter 'Wt mg' in the *Y-column* and 'Pip No' in the *X-column*;

2) Check that *Data display* contains opposite *Item 1* the default entries of *Symbol* and *Graph*;

3) Click on ☐OK☐.

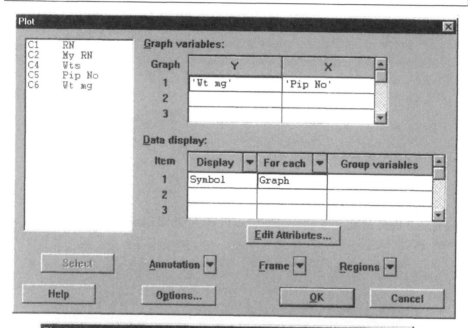

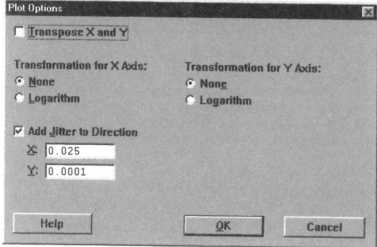

Figure 1.8   MINITAB dialog box for *Plot*. Top, main box; bottom, sub-dialog box for access to *Jitter* option

The computer should display the point plots of Figure 1.9, which shows (top) that Pipette 2 apparently has only two points because of coincidence. This can be got around with the *Jitter* option. So go back and do the plot again. This time, after Step 2 above:

---

1) Click on *Options*, and when the sub-dialog box opens (Figure 1.8, bottom);

2) Click on ☑ adjacent to *Add Jitter to Direction*;

3) Leave the default option of $\boxed{0.025}$ in the *X*, but change the *Y* to $\boxed{0.0001}$ (because no significant displacement in Y is wanted, and *Jitter* requires some number greater than zero);

4) Click on $\boxed{OK}$ to close the sub-box and on $\boxed{OK}$ in the main box.

This should deliver the altered point plot in Figure 1.9 (bottom), with *Jitter* giving slight horizontal displacements of all the points but still keeping them above the axis mark of each pipette.

The point plot suggests that the deliveries from Pipette 1 are, on average, greater than from the two other pipettes, especially No. 2, and also that the degree of scatter is larger with Pipette 3 than with Pipette 1. It is worth spending time with the *Plot* menu to explore its versatility. The *Jitter* function is only one of many options for altering the Graphics display.

In the next chapter another pictorial display, the histogram or bar chart, will be explored, and later the boxplot will be introduced in Chapter 3. They are not presented here as they are less useful with such small data sets as three groups of three observations each.

1.8 DESCRIPTIVE STATISTICS

1.8.1 Whole-group treatment

The nine weight readings can be treated either as a single set of observations, or as coming from three different pipettes. Both approaches may be valid according to the focus of interest. Thus it is reasonable to lump all the readings into a single set as representing what may happen in the laboratory when dispensing 100 µl volumes without taking account of which pipette is being used. On the other hand, to explore possible differences between individual pipettes, the data should be processed according to which pipette was being used. Initially the results will be taken as a single set of nine observations.

To get MINITAB to provide *Descriptive Statistics* of the nine observations, go to **Stat > Basic Statistics > Display Descriptive Statistics** and when the dialog box (Figure 1.10) appears:

---

1) In *Variables*, enter 'Wt mg'; this can either be typed in with the quotes, or entered by highlighting *Wt mg* in the left-hand listing of columns, and clicking on *Select*; alternatively the entry can be typed in as C6 (without quotes), but this is less useful for having on the final printout;

2) Click on $\boxed{OK}$.

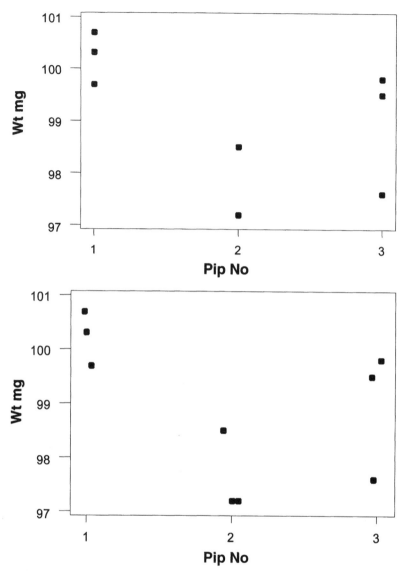

Figure 1.9 Point plot of the three weight results from each pipette. Top, default plot; bottom, with '*Jitter*' option (see text)

Observe the output in the Session window as in Table 1.1, where column numbers have been added for convenience of reference.

1.8.2 Explanations

MINITAB often provides much more output than may be wanted and the material of interest has to be selected from that provided. There are some items of the *descriptive statistics* that will be wanted every time, while others will hardly ever be needed. MINITAB presents them all. Note that each of the 11 items in the Session window (Table 1.1) is a descriptive statistic, i.e. a numerical value that describes a particular aspect of the data. This is what is meant by a *statistic*.

Table 1.1 Descriptive statistics for the nine pipetting deliveries treated as a single group of observations. Column numbers have been added for convenience of discussion

Col. 1 Variable	Col. 2 N	Col. 3 Mean	Col. 4 Median	Col. 5 Tr Mean	Col. 6 St. Dev.	Col. 7 SE Mean	Col. 8 Minimum	Col. 9 Maximum	Col. 10 Q1	Col. 11 Q3
Wt mg	9	98.944	99.500	98.944	1.352	0.451	97.200	100.700	97.400	100.050

Table 1.2 Descriptive statistics for the weights of the nine pipetting deliveries treated as three from each pipette. Column numbers have been added for convenience of discussion

Col. 1 Variable	Col. 2 Pip No	Col. 3 N	Col. 4 Mean	Col. 5 Median	Col. 6 Tr Mean	Col. 7 St Dev	Col. 8 SE Mean	Col. 9 Minimum	Col. 10 Maximum	Col. 11 Q1	Col. 12 Q3
Wt mg	1	3	100.23	100.30	100.23	0.50	0.29	99.70	100.70	99.70	100.70
	2	3	97.633	97.200	97.633	0.751	0.433	97.200	98.500	97.200	98.500
	3	3	98.967	99.500	98.967	1.193	0.689	97.600	99.800	97.600	99.800

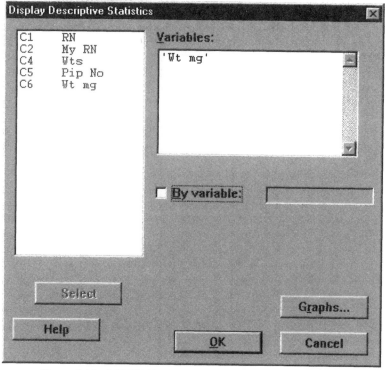

Figure 1.10 MINITAB dialog box for *Descriptive Statistics*

Taking the most useful items first and reading from left to right in Table 1.1:

Col. 1: Records that the *variable* being described statistically is *Wt mg* (unless C6 was entered as the column label);

Col. 2: Gives the *Number of Observations* as $N = 9$;

Col. 3: The *Mean* (also known as arithmetic mean) is the average, obtained by adding up the nine weights and dividing by nine;

The remaining items tend to be less frequently needed, some hardly at all.

Col. 4: The *Median* is the middle observation in the series when the values are arranged in numerical order from the lowest to the highest. If N were an even number, the median would be interpolated midway between the two middle values. The median is used as an alternative to the mean if the latter is not appropriate for some reason. It is also an indicator of the 'symmetry of the underlying distribution' (see Chapter 5).

Col. 5: The *Trimmed Mean* (Tr Mean) is rarely used in scientific publications. Its application is for very large data sets of, say, 50 or more observations. Sometimes such collections have a few exceptionally low or high 'rogue' values that distort the mean of the main bulk of the data. The trimmed mean is the mean that is calculated after discarding the top 5% and the bottom 5% of the observations, so that any exceptional values are excluded. In small data sets such as $N = 9$, the mean and the trimmed mean will be the same because there are not enough observations to discard the 5% from each end. For scientific work it is very doubtful whether data at the top and bottom ends of a set should be discarded in this way.

Col. 6: The *St. Dev.*, the *standard deviation*, is a measure of the extent of variation or scatter between the observations and is discussed in the next chapter (§2.3.4 and FN 2.1).

Col. 7: The *SE Mean*, the *standard error of the mean*, is a measure of the *reliability of the mean* and is also explained in the next chapter (§2.5.2).

Col 8: The *Minimum* is the smallest value in the set;

Col 9: The *Maximum* is likewise the highest value in the set;

Col 10: Q1 is the *First Quartile*. To get it, the data are arranged in numerical order from the lowest value to the highest value. Q1 is the boundary, interpolated as required, that defines the top edge of the bottom quarter (25%) of the data. Q2 (not shown as such) is the same as the median, or half-way up;

Col 11: Q3 is the *Third Quartile*, or the boundary between the lower three-quarters and the top quarter of the data. Thus the observations enclosed between Q1 and Q3 are the central 50% of the data in terms of numerical value, a grouping that is sometimes quite useful to see.

1.8.3 Each pipette separately

A very small alteration in the procedure used previously will give the *descriptive statistics* for each pipette separately, instead of treating the data as a single set of $N = 9$ observations. Go to **Stat > Basic Statistics > Display Descriptive Statistics**, and when the dialog box (Figure 1.10) appears:

---

1) Enter 'Wt mg' in $\boxed{Variables}$;

2) Click on $\boxed{\checkmark}$ beside $\boxed{By\ variable}$;

3) Enter 'Pip No' (with quotes) in $\boxed{By\ variable}$;

4) Click on $\boxed{OK}$.

Observe the output in the Session window as in Table 1.2. The layout is the same as before except for the extra Col. 2 for displaying the pipette numbers. Note the following:

1) The means of the three pipettes are different. Whether the differences are *significant* is a question for Chapter 3;
2) The means and the trimmed Means (Tr Means) for each pipette are identical, as expected for such small numbers of observations;
3). The st. devs vary quite a lot between pipettes; likewise the SE Means.
4). Cols 9 and 10 show that the *minimum* of Pipette 1 is higher than the *maximum* of Pipette 2. This suggests that Pipette 1 was delivering a larger volume, on average, than Pipette 2 (in conformity with the Point plot);
5) Pipette 1 and Pipette 3 only just overlap, again suggesting that Pipette 1 was delivering more, perhaps significantly more. This will be investigated in Chapter 3.

After finishing with MINITAB, go to the *File* window and *Save* the project. This saves the worksheet, the Session window and any graphs that may have been produced. Finally go to *Exit* to leave MINITAB.

1.9 SUMMARY

This chapter has emphasized the importance of collecting experimental observations with a process of randomization to minimize the confounding of variables, and has introduced MINITAB as a collection of procedures for statistical analysis. The statistical formulae used by MINITAB are not displayed by the computer. The pipetting data are intended as an example of the many different types of *measurement* (as distinct from *count*) data that are produced in the biological or biomedical laboratory. Note that some kinds of measurement data, such as antibody titres, may have to be transformed into logarithmic values before statistical analysis. This is discussed in Chapter 5.

FURTHER NOTES

FN 1.1 More about random numbers

In the example of the three pipettes, random numbers from 1 to 3 were needed. If the experiment had been with five pipettes, MINITAB would have been asked for random numbers from 1 to 5. Thus the range of random numbers needed is dictated by the number of groups in the experiment, not by the number of replications within each group.

As regards the *number* of random numbers, it is better to ask for many more than the total number of observations in the experiment because of the repetitiveness that is inherent in a random number series. That is why 20 random numbers were requested in an experiment with only nine observations. However, 200 random numbers could just as easily have been obtained.

For further insights into random numbers, take a look at a set of 400 random numbers from 1 to 3. These were generated as in Figure 1.3, except that the entry *C1–C20* was put in the *Store in column(s):* space. This gave a worksheet with 20

Table 1.3 A MINITAB-generated array of 400 of the random numbers 1, 2, and 3. Horizontal runs of four, five, and six of the same numbers are shown boxed in ⌐plain⌐ text, ⌐*italics*⌐, and ⌐**bold**⌐ respectively

```
3 1 1 1 1 2 2 3 1 1 3 2 3 1 2 2 2 2 2 1 2 2 2 3 2 3 3 1 1 2 1 2 3
3 1 2 1 3 3 1 3 2 3 2 2 1 1 1 2 2 2 2 3 3 3 1 1 3 3 1 2 2
3 2 1 3 2 2 1 2 2 3 3 3 1 2 3 1 3 2 2 3 3 2 1 2 2 2 1 3 1 3 1 2 1
1 2 3 1 1 1 2 3 3 1 1 2 2 2 3 1 3 1 1 3 1 2 3 3 3 3 2 3 2 2 1 1 1
1 2 2 2 1 3 3 1 1 3 2 1 2 3 2 2 2 1 2 2 2 1 1 3 3 3 3 3 2 2 2 2 3
1 1 3 3 1 1 2 3 3 2 2 1 2 2 3 2 3 1 1 2 3 2 2 1 3 1 2 3 1 2 3 3 3
2 1 2 2 2 2 1 2 2 3 2 3 1 1 3 2 3 2 3 3 2 2 2 2 2 1 2 2 1 2 2 1 3
3 2 1 1 1 3 2 1 2 2 1 1 1 3 3 3 2 2 2 2 2 2 1 1 2 3 2 3 2 1 3 2 3
2 1 3 3 1 3 1 2 3 1 3 2 3 1 3 1 1 1 3 2 2 1 1 3 1 3 1 2 2 1 2 3 1
3 3 2 1 1 3 3 3 2 3 3 3 2 2 1 1 1 2 2 1 1 3 2 2 2 3 1 1 3 2 3 3 1
2 2 2 2 1 1 2 1 1 1 1 1 2 1 1 3 2 1 3 1 1 1 3 1 2 3 2 3 2 3 2 3 3
1 3 1 1 3 3 1 1 3 3 1 1 3 3 2 3 1 1 3 2 1 1 2 1 3 1 3 2 3 1 2 3 1
2 3 1 1 2 1
```

rows and 20 columns that has been reorganized slightly for presentation in Table 1.3. In this case the random digits have been read horizontally (rather than vertically as in the pipetting experiment) and 'runs' of the same digit highlighted. This illustrates how difficult it would be to make up a random number series just by writing down numbers in what might look like randomness. Note that there are several runs of fours and fives of the same digit and one run of six of the same digit. Altogether the table contains 134 of the number 1, 142 of 2 and 124 of 3. These numbers are very close to a 1:1:1 ratio, as would be expected.

2

How to Condense the Bulkiness of Data

I often say that when you can measure what you are speaking about, and express it in numbers, you know something about it; but when you cannot measure it, when you cannot express it in numbers, your knowledge is of a meagre and unsatisfactory kind.

William Thompson [Lord Kelvin] (1824–1907)

2.1 EARLY STEPS

2.1.1 Journal tabulation versus MINITAB

In the last chapter a small collection of only nine observations was considered. Suppose now that the pipetting exercise, involving three deliveries from each of three pipettes, was performed by a group of 15 students. They each took turns (in a pre-determined random sequence!) at going to the balance and using the same three pipettes to collect a personal set of weighings of nine deliveries of 100 µl of water. This generated altogether a set of $9 \times 15 = 135$ observations for the whole group. A data set like this, if it were being published in a scientific journal, would probably be organized as in Table 2.1, which assigns one line to each student and nine columns for the collected observations, grouped by pipette. Apart from its size, this is a common type of data set, in having the two independent variables of *pipette* and *student*, and with replication.

This layout is not, however, suitable for MINITAB, which brings in a general point that the style of tabulating data that would be all right for a scientific journal may be unsatisfactory for MINITAB. Instead, MINITAB needs all the experimental measurements to be gathered into a single long column of *Wt mg*, and for the categorization by student and by pipette to be shown in code numbers in two adjacent columns alongside. Thus for MINITAB purposes the results in Table 2.1 should be set out as in Table 2.2, except that the layout should be as three long and uninterrupted columns, each with 135 lines. It is only to get the table onto the printed page that it has been chopped into three sections of three columns. Therefore to work through this exercise, the figures in Table 2.2 should be entered on to a MINITAB worksheet that will have the three columns (C1, C2 and C3) and 135 rows of Table 2.2. Note that only the *Wt mg* figures in C3 have to be copied, since the other two columns can be filled by the *Make Patterned Data* command below (and as previously used in §1.6.2).

Table 2.1 Conventional method (as for publication in a journal) of presenting the 135 pipette delivery weights from 15 students, each of whom made three deliveries from each of the same three pipettes

Student No.	Pipette 1			Pipette 2			Pipette 3		
1	98.0	96.9	98.9	97.7	97.1	96.6	97.1	96.0	95.7
2	100.8	98.7	99.3	98.2	98.2	97.6	99.8	99.5	99.5
3	99.8	100.1	100.0	99.1	99.9	99.0	101.9	99.6	99.5
4	99.6	99.4	100.0	99.7	100.0	103.2	97.7	99.3	100.2
5	103.3	99.4	99.0	98.8	96.7	99.6	101.2	101.3	99.8
6	100.1	100.0	99.8	100.2	99.5	99.5	102.2	102.1	101.7
7	99.0	100.1	97.4	98.1	95.8	95.1	98.1	98.0	98.1
8	103.8	101.0	101.3	103.6	100.8	101.7	102.3	102.3	101.2
9	100.9	101.3	101.6	97.9	98.2	98.1	99.4	100.0	100.0
10	97.4	97.0	96.9	96.3	96.3	97.2	99.9	99.4	100.3
11	99.1	99.1	99.1	99.5	99.2	99.9	99.1	99.4	99.1
12	95.3	97.4	97.4	95.7	98.4	100.8	99.8	99.8	100.7
13	100.7	99.4	101.2	100.0	99.6	97.5	100.4	98.1	99.2
14	100.1	99.8	98.8	100.4	98.8	99.5	102.6	101.3	102.6
15	97.2	99.1	96.8	95.8	95.9	100.1	97.1	95.9	96.1

Table 2.2 The students' 135 pipetting results entered into the first three columns of a MINITAB worksheet, but displayed here in a cut and spaced format so as to fit on the page

Stud No	Pip No	Wt mg	Stud No	Pip No	Wt mg	Stud No	Pip No	Wt mg
1	1	98.0	6	1	100.1	11	1	99.1
1	1	96.9	6	1	100.0	11	1	99.1
1	1	98.9	6	1	99.8	11	1	99.1
1	2	97.7	6	2	100.2	11	2	99.5
1	2	97.1	6	2	99.5	11	2	99.2
1	2	96.6	6	2	99.5	11	2	99.9
1	3	97.1	6	3	102.2	11	3	99.1
1	3	96.0	6	3	102.1	11	3	99.4
1	3	95.7	6	3	101.7	11	3	99.1
2	1	100.8	7	1	99.0	12	1	95.3
2	1	98.7	7	1	100.1	12	1	97.4
2	1	99.3	7	1	97.4	12	1	97.4
2	2	98.2	7	2	98.1	12	2	95.7
2	2	98.2	7	2	95.8	12	2	98.4
2	2	97.6	7	2	95.1	12	2	100.8
2	3	99.8	7	3	98.1	12	3	99.8
2	3	99.5	7	3	98.0	12	3	99.8
2	3	99.5	7	3	98.1	12	3	100.7
3	1	99.8	8	1	103.8	13	1	100.7
3	1	100.1	8	1	101.0	13	1	99.4
3	1	100.0	8	1	101.3	13	1	101.2
3	2	99.1	8	2	103.6	13	2	100.0
3	2	99.9	8	2	100.8	13	2	99.6
3	2	99.0	8	2	101.7	13	2	97.5
3	3	101.9	8	3	102.3	13	3	100.4
3	3	99.6	8	3	102.3	13	3	98.1
3	3	99.5	8	3	101.2	13	3	99.2
4	1	99.6	9	1	100.9	14	1	100.1
4	1	99.4	9	1	101.3	14	1	99.8
4	1	100.0	9	1	101.6	14	1	98.8
4	2	99.7	9	2	97.9	14	2	100.4
4	2	100.0	9	2	98.2	14	2	98.8
4	2	103.2	9	2	98.1	14	2	99.5
4	3	97.7	9	3	99.4	14	3	102.6
4	3	99.3	9	3	100.0	14	3	101.3
4	3	100.2	9	3	100.0	14	3	102.6
5	1	103.3	10	1	97.4	15	1	97.2
5	1	99.4	10	1	97.0	15	1	99.1
5	1	99.0	10	1	96.9	15	1	96.8
5	2	98.8	10	2	96.3	15	2	95.8
5	2	96.7	10	2	96.3	15	2	95.9
5	2	99.6	10	2	97.2	15	2	100.1
5	3	101.2	10	3	99.9	15	3	97.1
5	3	101.3	10	3	99.4	15	3	95.9
5	3	99.8	10	3	100.3	15	3	96.1

Note: Stud No = student number; Pip No = pipette number.

2.1.2 Making patterned data

To fill the entries for *Student Number*: Go to **Calc > Make Patterned Data > Simple Set of Numbers** and, when the dialog box appears, fill the sections as follows:

1) *Store patterned data in*: $\boxed{\text{Stud No}}$;
2) *From first value*: $\boxed{1}$
3) *To last value*: $\boxed{15}$
4) *In steps of*: $\boxed{1}$
5) *List each value*: $\boxed{9}$ *times*
6) *List the whole sequence*: $\boxed{1}$ *times*
7) $\boxed{OK}$.

To fill the entries for *Pipette Number*: Go to **Calc > Make Patterned Data > Simple Set of Numbers** and, when the dialog box appears, fill the sections as:

1) *Store patterned data in*: $\boxed{\text{Pip No}}$;
2) *From first value*: $\boxed{1}$
3) *To last value*: $\boxed{3}$
4) *In steps of*: $\boxed{1}$
5) *List each value*: $\boxed{3}$ *times*
6) *List the whole sequence*: $\boxed{15}$ *times*
7) $\boxed{OK}$.

This should recreate Table 2.2 on a MINITAB worksheet as three columns and 135 rows, and with the headings as shown.

2.2 HISTOGRAMS, OR BAR CHARTS

Before doing a detailed numerical analysis of any large set of measurement data, it is generally useful to get a preliminary visual impression as provided by a histogram, otherwise known as a bar chart. MINITAB, with all its options, offers a large number of different histogram variations, but to start only the automatic, or 'default', settings will be used.

2.2.1 Default settings

1) Go to **Graph > Histogram** and, when the dialog box (Figure 2.1) opens:
2) Highlight *C3 Wt mg* in the left-hand listing of the columns;
3) Move the cursor to the first row under X and opposite to Graph 1;
4) Press the $\boxed{Select}$ button to enter *Wt mg* at the cursor point opposite to Graph 1; note that it is entered with single quote marks, which means that for typing *Wt mg* into the box manually, single quotes would have to be added;
5) Click on $\boxed{OK}$.

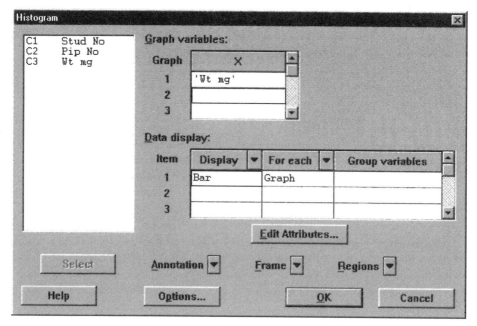

Figure 2.1   MINITAB dialog box for *Histogram*, with '*Wt mg*' entered as the variable

This gives the histogram in Figure 2.2, in which the observations have been sorted into 19 bars with a definite peak at around 100 mg. There is a suggestion that there might be three slightly separated peaks, rather than a single peak, but there is no strong asymmetry, or outliers. The degree of irregularity in the heights of adjacent bars is no more than is commonly seen in histograms of experimental data. If there were marked asymmetry, this would be a sign that the data might not be normally

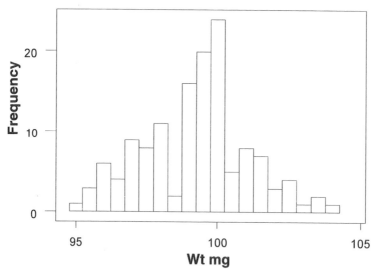

Figure 2.2   Default version of the histogram that MINITAB produced for the 135 pipetting results in Tables 2.1 and 2.2

distributed, and that the procedures in Chapter 5 should be applied. The histogram thus provides an initial pictorial impression of the data that would not be revealed by visual inspection of the rows and columns of the numbers in Tables 2.1 or 2.2. This is an example of the theme of this chapter, which is to explore some of the methods for condensing the bulkiness of data into a form that the human eye and mind can assimilate.

The vertical axis labelled *Frequency* refers to the number of weight observations that occurred within the width of each bar. The bar width is not stated explicitly on the histogram, but from the fact that there are 19 bars in a scale width of about 9 mg, each bar is about 0.5 mg wide. Thus the *Frequency* scale shows that the tallest bar contains about 24 observations that lie within 0.25 mg on either side of 100 mg.

2.2.2 Alternative presentations

MINITAB offers a large number of options for altering the appearance of histograms but only a few of these will be explored here. However, none is likely to reveal any major features that were not shown by the automatic option. But they will make the diagram look different. Among the most useful of these possibilities in rough order of importance are:

1) Preselecting the number of bars instead of accepting the automatic option. Note that the automatic option is not necessarily 19 bars as here, but depends on the size of the data set. MINITAB allows choice of any number of bars between 2 and 99;
2) Having the vertical axis scale in *Percent* instead of *Frequency*. Thus each bar contains a displayed percentage of the total number of observations, instead of the actual number of observations. This may be useful for bringing data sets of different sizes to a common percentage scale for comparison purposes;
3) Preselecting a particular bar width;
4) Preselecting whether each bar is plotted above its midpoint as in Figure 2.2, or with the axis ticks coincident with the edge of each bar, what MINITAB calls the *cutpoint* option;
4) Having the bars shaded or dotted, or filled with a solid colour, instead of being left white, as in Figure 2.2. Likewise the background can be coloured, or left plain;
5) Other artwork changes such as altering the thickness of the lines, and the size and types of font, and adding legends, etc., as required;
6) Having several histograms asssembled together by the process of *Tiling*. Figures 2.3 to 2.6 show a small selection from the many choices available.

Figure 2.3 involved two changes from the default settings: having entered *Wt mg* as the graph variable on the opening dialog box as before, the *Option* button was clicked to open an internal dialog box. Here, *Percent* was selected instead of *Frequency*, and the *Number of intervals* was clicked so as to allow the entry of 6. This considerably altered the appearance of the histogram with seemingly a single peak slightly displaced to the left of that in Figure 2.2, but still reasonably symmetrical.

In Figure 2.4, the number of bars was set at 10, although the histogram seems to show only eight because the end ones are so small. In addition the axes were turned through 90° by activating the *Transpose x and y* button. Also the *Edit* button was

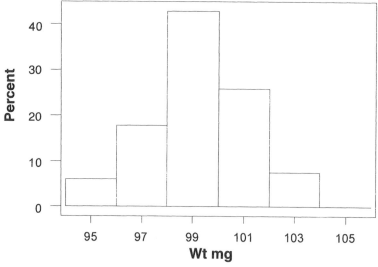

Figure 2.3   Alternative histogram of the 135 pipetting results with the bar number set at 6 and the ordinate axis altered to *Percent*

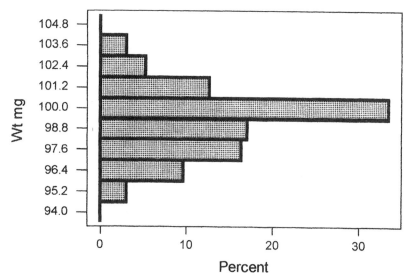

Figure 2.4   Alternative histogram of the 135 pipetting results, with the bar number set at 10, the *X*- and *Y*-axes transposed and the bars filled with a pattern of squares

pressed to open a third window which allowed more changes from the default settings: *Fill type* was entered with *Squares*; and the lines were made more prominent by selecting a line thickness of 3 in *Edge size*. This histogram shows a very sharp peak centred on 100 mg as in the first histogram.

In Figure 2.5, the intervals were kept at 10 and the *Transpose x and y* feature deactivated. However the y-axis was labelled in *Density* which presents each bar as a decimal fraction of one. In the *Edit attributes* box, the *Fill Type* was changed to *Right slant* and the *Edge size* kept at 3.

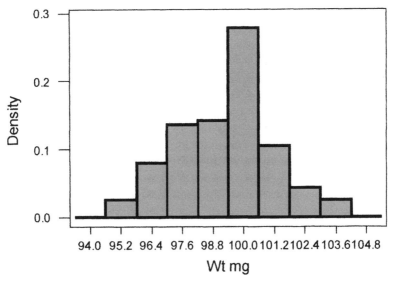

Figure 2.5 Alternative histogram of the 135 pipetting results, with the bar number set at 10, the axes returned to normal setting, the *Y*-axis expressed in *Density* and the fill pattern changed to *Right slant*

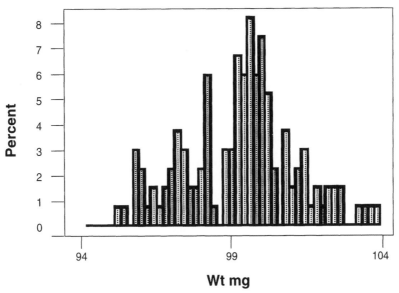

Figure 2.6 Alternative histogram of the 135 pipetting results, with the bar number set at 50, the *Y*-axis in *Percent* and the *Dots* option chosen as the *Fill* pattern

In Figure 2.6, The default option of *Percent* was chosen for the vertical axis but the number of bars was set at 50 in the *No of intervals* box. From the *Edit* menu, the *Dots* option was selected from *Fill Type* and the *Edge Size* increased to 3 from the default option of 1. Note that this histogram agrees with the impression suggested by the first one, where there were 19 bars, that the data might consist of three slightly displaced peaks, but again without any suggestion of asymmetry. Thus while one should not try to read too much from histograms, they may give clues as to what may become consolidated during more detailed analysis.

2.3 FROM HISTOGRAM TO NORMAL DISTRIBUTION

Most readers will be familiar with the normal, or Gaussian, distribution curve – the bell-shaped curve that represents many types of measurement data. MINITAB has provision for superimposing the appropriate normal curve onto a histogram and, at the same time, providing the descriptive statistics of mean, st. dev., etc.

2.3.1 Descriptive statistics and normal distribution curve

1) Go to **Statistics > Basic Statistics > Display Descriptive Statistics** and, when the dialog box opens:

2) Highlight *C3 Wt mg* in the box of column listings;

3) Press the $\boxed{Select}$ button to enter '*Wt mg*' (note the addition of single quotes) in the *Variables* box;

4) Click on the *Graphs* button and, when the internal window opens, select *Histogram of data with normal curve*;

5) Click on $\boxed{OK}$ to close the *Graphs* window;

6) Click on $\boxed{OK}$ again to close the main window.

These steps deliver the *Descriptive Statistics* in the *Session* window and, in front of that, a graph with the same 19 bar histogram as in Figure 2.2 but with the normal distribution curve superimposed. This is Figure 2.7, which shows a histogram whose outline only very approximately fits the superimposed bell-shaped curve.

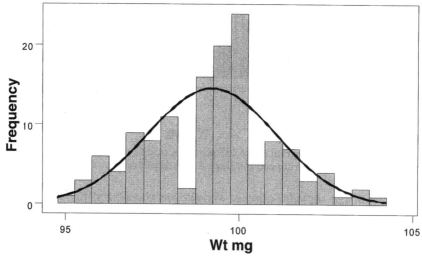

Histogram of Wt mg, with normal curve

Figure 2.7 Default histogram and superimposed normal distribution curve delivered by MINITAB on the 135 pipetting results, from the *Display Descriptive Statistics* command

From the *Descriptive Statistics*, the two most important items are selected, namely the Mean = 99.233 and the St. dev. = 1.85.

Variable	N	Mean	Median	Tr Mean	St. dev.	SE Mean
Wt (mg)	135	99.233	99.400	99.221	1.850	0.159

Variable	Minimum	Maximum	Q1	Q3
Wt (mg)	95.100	103.800	98.000	100.100

2.3.2 Diversionary exercise with a much larger data set

Here, the values mean = 99.2 and st. dev. = 1.85 are taken from the 135 student results and, purely as an illustrative exercise, entered into MINITAB to generate a much larger set of *fictitious* pipetting results from a true normal distribution with Mean = 99.2 and St. dev. = 1.85. Instead of 135 results, MINITAB was asked to produce 5000, or 37 times as many.

First open a new MINITAB folder and worksheet, and label the first column *Wt mg* to receive the fictitious set of 5000 pipetting results. Then go to **Calc > Random Data > Normal** and, when the dialog box opens:

> 1) Highlight *C1 Wt mg* in the box of column listings;
> 2) Enter 5000 in the box bounded by *Generate* and *rows of data*;
> 3) Enter 99.2 opposite to Mean;
> 4) Enter 1.85 opposite to *St. dev.*;
> 5) Click on $\boxed{OK}$.

Scrolling down the worksheet will reveal that this has produced a long thin column of 5000 values of fictitious *Wt mg* readings. These were then processed as the 135 real results above, to give descriptive statistics in the Session window and a histogram with superimposed normal distribution curve as in Figure 2.8. Note that this histogram has a large number of very narrow bars and is a much better fit to the normal distribution curve than were the 135 actual results. Why was 5000 chosen? Because it is about as large a set as MINITAB can manage despite the worksheet having 100 000 cells. For example, when 13 500 randomly distributed normal values were requested in C1 (i.e. to give a data set 100 times larger than the 135 student results), MINITAB was not able subsequently to process them into a histogram with a superimposed normal curve.

2.3.3 Infinitely large data set

The next step in the argument is to imagine that the data set, instead of having 'only' 5000 observations, has an infinitely large number. Under these conditions the histogram bars would be infinitely thin and infinitely numerous, like the pages of a book seen end on, and the minor unevenness at the top outline would have been ironed out into the smooth curve of the normal distribution. This, of course, is a purely theoretical idea to which real-life data can only approximate to a greater or lesser extent. A large part of statistical methodology is based on the assumption

Histogram of Wt mg, with normal curve

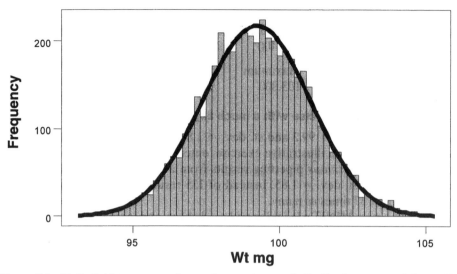

Figure 2.8 Default histogram, and superimposed normal distribution curve, delivered by MINITAB on a set of 5000 normally distributed random data with the same mean and st. dev. as the 135 pipetting results

that many types of real-life measurements *at least approximate* to a normal distribution. Therefore it is important to have objective tests to examine the validity of such an assumption.

One can think of chemistry as providing an analogy. When a research chemist buys a bottle of 'pure' (analytical grade) sodium chloride, this does not mean that there is not a single atom of, say, potassium in the bottle. All the manufacturer can do is to issue a certificate of purity stating that potassium and other potential impurities are below certain limits of detection. For most practical purposes the sodium chloride can be regarded as being 'pure'. Indeed for certain purposes a measurable amount of impurity may not be particularly critical. It is the same in statistics with measurement data and the normal distribution. For statistical purposes, the real-life data do not have to be an *exact* fit to a normal distribution to allow valid treatment as 'normally-distributed' values or to be described as 'normal'. Nevertheless it is important to run checks on closeness-of-fit to a normal distribution in order to avoid possible wrong conclusions (see §3.4).

2.3.4 Meaning of standard deviation

It always surprises me what a high proportion of students who take courses in statistics are unable, one year later, to give a reasonable answer to the question 'What is standard deviation?' I suppose that after the exam, when the mental dialog box gave the message 'Do you wish to save this file?' they pressed the 'No' button. It would seem that whatever they were taught did not leave any lasting impression.

I believe the best way to understand st. dev. is through a diagram such as Figure 2.9 with a normal distribution curve superimposed on the histogram of the class

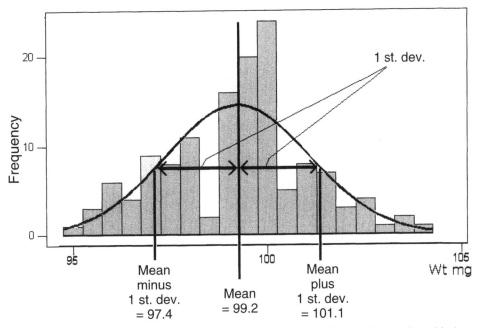

Figure 2.9 Histogram and normal distribution curve of the 135 pipetting results, with the standard deviation depicted as the horizontal distances from the mean to the points of inflection of the normal curve

pipetting results. The st. dev. is the distance from the mean line out to the sides of the Normal curve so as to meet it at the points where the slope changes from getting steeper to getting shallower (on the left side, and vice versa on the right, i.e. the *points of inflection*). Thus on the left-hand side, if one imagines the curve as the cross-section through a symmetrical mountain that starts out at minus infinity, where the slope is zero (horizontal). Moving rightwards, the slope then gets progressively steeper up to the st. dev. intersection point about half-way up, after which it becomes shallower. At the very top the slope is zero again, where the mean line intersects. Then the pattern is mirrored down the right-hand side, but with the slope becoming steeper as far as the point of intersection with the st. dev., beyond which it gets shallower and becomes horizontal at plus infinity.

2.4 AREAS UNDER A NORMAL DISTRIBUTION CURVE

2.4.1 Mean ± 1 st. dev.

The entire bell-shaped area under a normal distribution curve, down to the horizontal axis, corresponds to 100% of the data. Likewise the mean line divides the data into a lower 50% and an upper 50%. As shown in Figure 2.10, the area enclosed by ± 1.0 st. dev. on each side of the mean of a normal distribution curve when connected down to the horizontal axis encloses 68.2% of the data. This of course applies exactly only to a sufficiently large set of data that is a close approximation to a normal distribution. To the extent that there is some departure from normality, then the figure of 68.2% may be correspondingly altered.

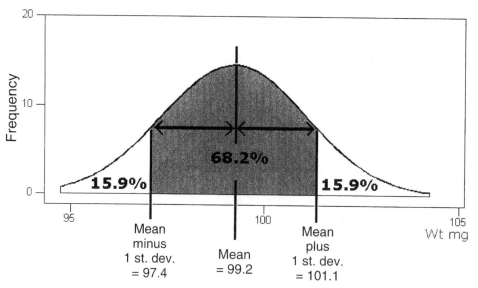

Figure 2.10 Normal distribution curve from the 135 pipetting results, with the area enclosed by ± 1 st. dev. around the mean being 68.2% of that under the curve

Therefore a useful check on the reliability of a st. dev. value from real-life data is to see whether ± 1 st. dev. around the mean does in fact enclose 68% of the observations. With the 135 student pipetting results, ± 1 st. dev. around the mean would be from 99.233 – 1.850 = 97.383, to 99.233 + 1.850 = 101.083.

The 135 results can now be classified as being either within or outside these limits. This could be done manually, but rather laboriously, from Table 2.2 by putting a pencil tick beside each result that is within the limits and then adding up all the ticks. The expected number would be 68.2% of 135 = 92.07, i.e. 92 for practical purposes. The same thing can be done on MINITAB with a formula that assigns the code number 0 to results outside the limits and the code number 1 to results that lie within the limits. To do this, open the worksheet with the 135 pipetting results and label two new columns, C5 with *Enclosed* and C6 with *Tally 1* (Table 2.3). Then go to **Calc > Calculator** and, when the dialog box opens:

---

1) Place the cursor in the box labelled *Store result in variable*;
2) Highlight *C5 Enclosed* in the box of column listings and press the *Select* button, so that it is transferred to the *Store result in variable* box;
3) In the *Expression* box, compose the formula '*Wt mg*'<101.083And'*Wt mg*'>97.383. Note that 'Wt mg' can be entered in the *Expression* box by highlighting *C3 Wt mg* in the Columns box and pressing the *Select* button. Note also how MINITAB attached single quotes to *Wt mg* in making the transfer;
4) Click on OK.

Table 2.3 The first 20 rows of the MINITAB worksheet of the students' 135 pipetting results, showing the use of the *Tally* function to determine the number of results that were within ± 1 st. dev. of the mean. Col. 4 was a blank, spacer, column and is not shown

Col. 1	Col. 2	Col. 3	Col. 5	Col. 6
Stud No	Pip No	Wt mg	Enclosed	Tally 1
1	1	98.0	1	91
1	1	96.9	0	
1	1	98.9	1	
1	2	97.7	1	
1	2	97.1	0	
1	2	96.6	0	
1	3	97.1	0	
1	3	96.0	0	
1	3	95.7	0	
2	1	100.8	1	
2	1	98.7	1	
2	1	99.3	1	
2	2	98.2	1	
2	2	98.2	1	
2	2	97.6	1	
2	3	99.8	1	
2	3	99.5	1	
2	3	99.5	1	
3	1	99.8	1	
3	1	100.1	1	

Note: Stud No = student number; Pip No = pipette number

Inspection of Col. 5 (Table 2.3) shows it now filled with zeros and ones, according to whether the *Wt mg* value was inside or excluded from the ± 1 st. dev. limits.

To add up the number of 'ones', go to **Calc > Calculator** and, when the dialog box opens:

1) Place the cursor in the section labelled *Store result in variable*;

2) Highlight *Tally 1* in the box of column listings and press the *Select* button, so that it is transferred to *Store result in variable*;

3) Transfer the cursor to the *Expression* box and then scroll down the options in the *Functions* box until the entry *Sum* is found. Highlight this and press *Select* to transfer it to the *Expression* box, where it appears as *Sum(number)*;

4) Replace *number* with highlighted C5 *Enclosed* and pressing *Select*. The *Expression* box should now contain *Sum('Enclosed')*;

5) Click on ⎡*OK*⎤.

In the column headed *Tally 1* (Table 2.3) there is now the single entry of 91. This is the number of '1'values in the *Enclosed* column and is very close to the expected figure of 92. This outcome provides strong support for the view that the 135 student pipetting results are close to being 'normally distributed', or 'normal' for short, despite the apparent irregularities in the various histograms.

2.4.2 Mean ± 1.96 st. dev.

Another accepted way of dividing up the area under a normal distribution curve is that in Figure 2.11 where the central area enclosing 95% of the data is shaded. This figure of 95% is a standard benchmark quantity in statistics and is taken as representing the vast majority of the observations in a data set, i.e. 95% of them, or 19 out of 20.

This is comparable to a court of law where a conclusion is reached 'beyond reasonable doubt'. A statistical conclusion that is consistent with 95% of the data can be taken as being 'beyond reasonable doubt'. It does not mean that there is *no* possibility of a mistake or misinterpretation, but since it embraces the over-whelming mass (95%) of the data, it constitutes a reasonable conclusion from the evidence available. Loopholes for exceptions are provided by the 2.5% of the observations that are below the broad central 95% zone and the 2.5% that lie above it. These two small areas are referred to as the lower and upper 'tails' of the distribution. In order to have exactly 95% of the area enclosed, the distance on either side of the mean has to be 1.96 st. dev., and is the explanation for what may seem a peculiar number.

With the 135 student pipetting results, the limits of mean ± 1.96 st. dev. are 95.607 and 102.859. These are rounded to 95.6 and 102.9 in Figure 2.11 but, for what follows, it is best if the extra decimal places are retained, to avoid placement ambiguities in determining which data points are inside and outside the limits. With 135 observations the limits of ± 1.96 st. dev. about the mean should contain 128 of the observations if the data are normally distributed.

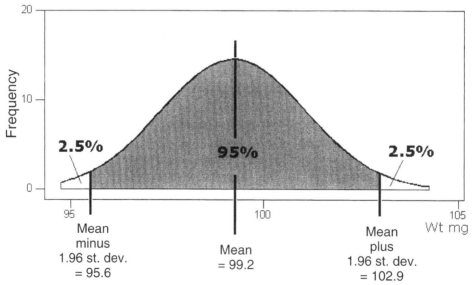

Figure 2.11   Normal distribution curve from the 135 pipetting results, with the area enclosed by ± 1.96 st. dev. around the mean being 95% of that under the curve, leaving two tails each with 2.5%

This can be explored by the same procedure as used previously. Col. 7 on the Worksheet can be labelled *1.96 SD* and Col. 8 labelled *Tally 2*. The formula in the *Expression* box, should be '*Wt mg*'<102.859And'*Wt mg*'>95.607. At the end of the operations, the figure in *Tally 2* is 129, which is very close to the value of 128 that was expected from the assumption of normality of the distribution. This provides further support for the conclusion that the 135 pipetting results are a reasonably good fit to a normal, or Gaussian, distribution.

2.5 SAMPLE AND POPULATION

2.5.1 Thinking statistically

Going back to the data in Chapter 1, there were three replicate observations from each of three pipettes. In statistical thinking, such a set of three replicate weighings of water from an individual pipette is to be regarded in a special way. It is to be thought of as a *sample of size N = 3* taken from a *population of size N = infinity*. That is, although the investigator collected only three observations, there was no reason for stopping at three.

Hypothetically, the investigator and his/her children and grandchildren and the rest of mankind could have continued making weight observations with that pipette until the end of time, and thereby build up an almost infinitely large set of weight observations. But life being short, the decision was taken to limit the sample size to $N = 3$. Nevertheless and theoretically, there is 'out there' in the statistician's imagination an infinitely large set of replicate weighings that in principle *could be taken* from a particular pipette. This infinitely large set is known as the population. It has no actual physical existence as a real data set, but it has to be taken into account to explain the *SE* mean and *95% confidence limits* that are discussed below.

So in statistical terms the $N = 3$ actual observations should be thought of as a sample from a theoretical and infinitely large population of pipetting weights that has no physical existence. Thus the mean and st. dev. of the sample are *estimates* of the mean and st. dev. of the 'underlying' population (that exists only as a statistical idea). If by some magical process it was possible to get the exact value of mean and st. dev. of the population (of weight readings of that student and that pipette), then the 'truth' would be revealed. In reality, however, it is only possible to estimate the values of the population mean and st. dev. from the values from the sample of $N = 3$ observations. The fact that the three observations were not identical means that if the same student were to collect further sets of $N = 3$ observations from that pipette, they would probably not have exactly the same values of mean and st. dev. as the first set.

So the word population in statistics has a special meaning. In the early days of the subject it was applied to the health and wealth data from the population of the people in a country. But since then, the meaning of 'population' has become broader. In experimental science, population refers to the usually infinitely large data set that has no physical existence as such, but from which one or more samples (the experimental observations) were taken.

The infinity feature for a population is not obligatory. For example, in public opinion polls, the population may be that of the country or relevant group therein, from which a sample of individuals was selected for questioning. And just as in experimental science, there is the assumption that the sample, which may be of size $N = 1000$, gives a true picture of the 'underlying' population, which may have $N = 50$ million. All this ties in with the need to avoid bias in obtaining the sample, which is why the process of random sampling is so important in the conduct of experiments and surveys.

Thus the purpose of many statistical methods is, by a process of inference, to use observations on samples to obtain information on populations. The degree of reliability will be determined by the inherent variability of the individuals in the population (be they people, or pipette delivery weights), the quality and size of the sampling and the use of appropriate statistical methods. The art of statistics, as applied to experimental science, is to choose appropriate numbers of replicates (i.e. N-values) for the level of precision needed to answer a particular question, and taking account of the inherent variability in the 'population'.

2.5.2　Standard error of the mean

The *standard deviation (st. dev.)* and the *standard error of the mean (SE mean)* are related to each other, but have different meanings and are frequently confused. The above discussion showed that st. dev. can be visualized as the distance out to the points of inflection on the normal distribution curve. Also, about 68% of the observations in the sample are enclosed within the limits of ± 1 st. dev. of the mean, as represented by the shaded area under the curve in Figure 2.10.

SE mean, on the other hand, is easy to define but not so easy to visualize in the mind's eye. The definition is:

$$\text{SE mean} = (\text{st. dev.})/\sqrt{N} \qquad\qquad \text{Eq. 2.1}$$

Thus with the 135 pipetting results where MINITAB gave st. dev. $= 1.850$ and SE mean $= 0.159$, we have SE mean $= 1.850/\sqrt{135} = 1.850/11.62 = 0.159$, as given in §2.3.1 in the *Descriptive Statistics*.

To understand SE mean, it is necessary to remember that the $N = 135$ observations are a sample from a hypothetical 'underlying' population of pipette weighings where $N =$ infinity. Thus putting $N =$ infinity into Eq. 2.1 makes the SE mean become zero. This is because the square root of infinity is still infinity, and any number divided by infinity is zero. Under these conditions the exact value of the population mean would be known, and without any uncertainty associated with it. For that reason the SE mean would be zero. At the same time the st. dev. would *not* be zero but would be close to the st. dev. $= 1.85$ of the 135 pipetting results and would be the actual population st. dev., instead of just being an estimate of it.

For a diagrammatic presentation, the SE mean can be shown as bars projecting horizontally on either side of the mean (Figure 2.12). In biological research papers where graph points commonly have SE mean (SEM) bars attached, they would generally be longer than here (in relation to the mean) because there would rarely be as many as $N = 135$ observations going into each graph point. Therefore the st. dev. would rarely be diminished by a divisor as large as the square root of 135. Also the st. dev. itself would probably be larger than here, because whatever was being measured in a biological experiment would likely have more variability than the pipetting of water.

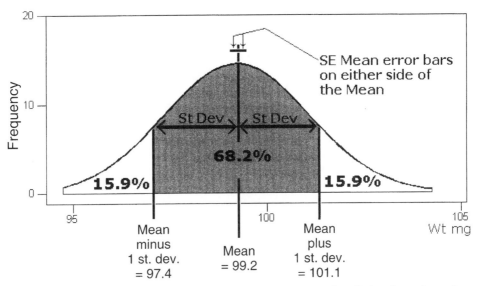

Figure 2.12 Standard error of the mean (SE mean) shown as being distinct from the st. dev. on a diagram of the normal distribution curve from the 135 pipetting results

For purposes of reporting, a statistical summary of the 135 pipetting results would commonly include the mean, the SE mean and the number of observations. So for the 135 pipetting results the summary statement would be: mean weight (mg) of 100 μl deliveries of water from three pipettes by 15 students:

$$\text{mean} \pm \text{SE mean } (N) = 99.233 \pm 0.159 \ (135).$$

2.5.3 Ninety-five percent confidence intervals

The mean produced by MINITAB from a set of experimental data should be regarded as a sample mean. It *estimates* the 'underlying' population mean whose true and exact value is rarely (if ever) known with complete finality. By the same reasoning, the st. dev. produced by MINITAB is the sample st. dev., which in turn is an *estimate* of the st. dev. of the 'underlying' infinitely large population of hypothetical pipette observations that only exists as an idea. The SE mean provides a reliability figure for attachment to the sample mean but it does not say how close it is to the population mean. However, it is possible to go one step further by obtaining the 95% confidence intervals (CI), also known as 'confidence limits'. This is not part of the information that MINITAB provides under *Descriptive Statistics* but can be obtained as follows in the MINITAB worksheet with the 135 pipetting results:

Go to **Stat > Basic Statistics > 1-Sample t** and, when the dialog box opens:

1) Highlight *Wt mg* in the left-hand listing of the columns;
2) Press the $\boxed{Select}$ button to enter *Wt mg* in the *Variables* box;
3) Click on $\boxed{OK}$.

The output then appears in the Session window as:

Variable	N	Mean	St. dev.	SE Mean	95.0% CI
Wt (mg)	135	99.233	1.850	0.159	(98.918, 99.547)

The first five columns repeat part of the output of descriptive statistics, but at the right-hand end are the 95% CI. These are the intervals, or limits, within which there is a 95% probability that the true population mean is actually located. So although the *exact* value of the population mean is not revealed, statistical theory and methods allow a probability statement to be made about it. It can be bracketed as having a high (95%) probability of being within the confines of the 95% CI, i.e. between 98.92 and 99.55 mg.

Note that the 95% CIs as given above are wider than the mean ± 1 SE mean by a factor of about 2. This factor, known technically as 't', is variable, depending on the value of N, but is never less than 1.96 and may be considerably larger. Thus in this example the 95% CI are ± 0.315 on either side of the mean instead of ± 0.159, for the SE mean. Probably because of their greater width, it is commoner for published results to use SE mean rather than 95% CI, as error bars on graphs, because the CI bars are at least twice as large and make the results look less exact. But in a sense the CIs are better than SE means because they make a direct probability statement about the limits within which the true population mean is likely to be located.

2.6   SUMMARY

This chapter has used as an example a set of 135 results produced by a group of 15 students working with three pipettes from which the weights of triplicated 100 µl deliveries of water were obtained. The data set was treated as a single group of $N =$ 135 observations for delivery of descriptive statistics. It is for a later chapter (§4.1) to investigate the 'internal anatomy' of the data, i.e. the variation attributable to the different pipettes and to the different students. After pictorial representation of the observations as histograms, the weight readings were used to illustrate the characteristics of the normal distribution, which the data fitted quite closely. The terms st. dev. (standard deviation), SE mean (standard error of the mean) and 95% CI (confidence intervals) were explained. Several useful MINITAB procedures were introduced. If the data had turned out not to be normally distributed, the procedures described in Chapter 5 would have been used.

FURTHER NOTES

FN 2.1   Standard deviation

Standard deviation, which in MINITAB is abbreviated to st. dev., is usually given the symbol 's' and is defined by the equation:

$$s = \sqrt{\frac{\Sigma \, (X - \bar{X})^2}{(N - 1)}} \qquad \text{Eq. 2.2}$$

To illustrate its use with small numbers, let us suppose there are two ($N = 2$) observations (X_1 and X_2) for which the st. dev. is wanted and their values are $X_1 =$ 2 and $X_2 = 3$. Bear in mind that from a statistical standpoint, the fact that standard deviation is being calculated implies a belief that the two pieces of data form a sample from an underlying normal distribution (otherwise there would be no point in calculating st. dev. as a *measure of dispersion*).

The *mean* $(\bar{X})$ is defined by:

$$\bar{X} = (X_1 + X_2 + \ldots\ldots X_n)/N \qquad \text{Eq. 2.3}$$
$$= (2+3)/2 = 2.5.$$

The *deviations from the mean* $(X - \bar{X})$ are:

For X_1: $(2 - 2.5) = -0.5$, and for X_2: $3.0 - 2.5 = 0.5$

The *squares of the deviations from the mean* are -0.5^2 and $0.5^2 = 0.25$ and 0.25

$\Sigma (X - \bar{X})^2$ is the *sum of the squares of the deviations* $= 0.25+0.25 = 0.5$

$(N - 1)$ is the *degrees of freedom* and is equal to *one less than the number of observations*. Here $(N - 1) = 2 - 1 = 1$.

Degrees of freedom (d.f.) is an idea deeply rooted in statistical theory and is concerned with the number of independent pieces of information contained in a sample and which can be used to obtain valid information about the 'underlying' population. Here the sample had two independent values 2 and 3, but as soon as $(\bar{X}) = 2.5$ had been calculated, then only *one* independent value, either the 2 or the 3, was left. This is because after the mean has been calculated, and either the 2 or the 3 selected as an independent observation, the other observation is automatically known and is no longer independent. Therefore to find the mean of $\Sigma(X - \bar{X})^2$, in spite of it containing the squares of the deviations of *two* observations, the correct divisor is 1 and not 2. This can be seen to make sense since two pieces of data can only provide *one* measure of dispersion.

FN 2.2   Variance

When the *sum of squares* $[\Sigma(X - \bar{X})^2]$ is divided by the degrees of freedom $(N - 1)$, i.e. $0.5/1 = 1$, the result goes under two names, the *variance* and also the *mean square*. The square root of the *variance* is the st. dev., i.e.:

$$\text{st. dev.} = \sqrt{variance} \qquad \text{Eq. 2.4}$$

In this case, $s = \sqrt{0.5/1} = 0.71$.

The procedure (Chapter 3) for investigating the differences in the mean deliveries of fluid from the three pipettes is analysis of variance, or ANOVA. The terms sum of squares, degrees of freedom and mean square, or variance are central to the ANOVA procedure. The degreees of freedom concept permeates the whole of statistical thinking. It is an essential feature of the *t-test*, *F-test* and *chi-square test* and also in the calculation of *confidence intervals*.

FN 2.3   Standard deviation and normal distribution

The top line of the st. dev. equation is generally written as $\Sigma(X - \bar{X})^2$, but this is not how either the computer or an electronic calculator works with it. Instead, use is made of the algebraic identity:

$$\Sigma(X - \bar{X})^2 = \Sigma(X^2) - (\Sigma X)^2/N \qquad \text{Eq. 2.5}$$

With the values of $X = 2$ and 3 from above:

$$\Sigma(X^2) + (\Sigma X)^2/N = 2^2 + 3^2 - (2 + 3)^2/N$$
$$= 4 + 9 - 25/2$$
$$= 0.5$$

which is the same as given by the other version of the formula. The benefit of using this second version is that the calculator first sets aside two of its memories. Then as each value of $\bar{X}$ is entered, one of the memories accumulates the values of X^2 while the other collects the values of X, and squares the total of the X-values when all the data have been entered.

A separate point about st. dev. is that the value of 's' delivered by Eq. 2.2 is the st. dev. of the sample, otherwise known as the sample standard deviation. It is to be distinguished from the 'underlying' population standard deviation, which is given the Greek letter σ, and of which s is an estimate. In the same way, the mean of the sample ($\bar{X}$) is an estimate of the 'underlying' population mean (μ).

The larger the sample size, the more exact is the knowledge of σ and μ. If there was an infinitely large sample, σ and μ would be known exactly, but in real life the best that can be done is to estimate them through the values of $\bar{X}$ and s, and attach confidence intervals to indicate the inexactitude of the knowledge. When MINITAB fits a smooth normal distribution curve on top of a histogram, it uses the $\bar{X}$ and s of the data as if they were μ and σ. The equation for the normal distribution may be found in standard texts and is not reproduced here.

FN 2.4 Precision and accuracy

These two terms are often used loosely and incorrectly. They have different meanings. *Accuracy* contains the idea of agreeing with an accepted reference measurement or 'target' value, whereas *precision* is concerned with closeness of reproducibility in replicate observations. Figure 2.13 uses target shooting to illustrate these points. In Target A, where all the shots are tightly grouped within the bull's-eye, both precision and accuracy are high. In Target B, the precision is just as high because the shots are in an equally tight group, but the accuracy is lower. This could result from some fault in the rifle barrel or badly adjusted sights, or because of a steady crosswind or some consistent fault in aiming. Any of these factors could affect accuracy without necessarily reducing precision. The corresponding situation in the pipetting experiment might arise if a pipette had a fault in calibration, or there was failure to expel the last drop of fluid in the tip. Target C shows reasonably accurate shooting, in that the shots are centred around the bull's-eye, but the precision is not as good as in A. Such shooting would be expected from an accurate rifle being fired by someone with shaky aim. In pipetting, similar variation could arise with an accurate pipette being used with a viscous fluid, and the operator not allowing adequate time for drainage of each delivery. In target D, both precision and accuracy are poor, as indicated by the scattered shots well away from the centre.

In the pipetting, standard deviation (st. dev.) is the measure of precision. A small st. dev. corresponds to a 'tight group' and a large st. dev. to widely scattered shots. Precision in the pipetting can be considered either with individual pipettes or with

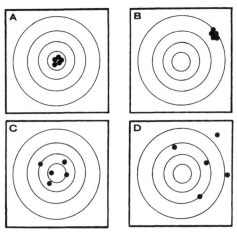

Figure 2.13   Precision and accuracy illustrated by target shooting: A, precision and accuracy both very high; B, precision as high as in A, but accuracy low; C, quite accurate, but precision lower than A or B; D both precision and accuracy are low

the nine readings as a whole. Precision may be measured as the *coefficient of variation (CV)*. This is defined as the st. dev. divided by the mean and expressed as a percentage:

$$CV = 100s/\bar{X}$$ Eq. 2.6

Thus with the nine weight readings taken together:

$$CV = 100 \, (1.352/98.944) = 1.4\%$$

Thus in a test with three pipettes, 100 µl water was delivered with a consistency, expressed as the CV, of 1.4%. Since the limits of ± 1 st. dev. on either side of the mean should enclose about 68% of the observations in the group, the CV of 1.4% defines the boundaries within which 68% of the observations should lie. It also implies that about 32% of the observations are outside the limits of ± 1.4% of the mean.

Turning now to *accuracy*, in the context of shooting, an accurate shot is one that hits the bull's-eye. But one could ask for a still higher standard of accuracy, by insisting that the first shot should hit the geometric dead-centre of the bull's-eye and that each successive shot should pass through this first hole without widening it. Most people would consider this unreasonable. So what is generally called 'accurate shooting' is dictated by the diameter of the bull's-eye and whether all the shots hit it.

In pipetting, the analogy to the area of the bull's-eye is the width of the *tolerance limits*, and like the size of the bull's-eye, these are arbitrary. For example, if *tolerance limits of accuracy* are set at 1%, this would be a weight range of 1% around 99.82 mg, the true weight of 100 µl H_2O. This range is:

1% tolerance limits = 99.82 ± 0.9982 mg = 98.82 to 100.82 mg

Taking the observations in the order they are listed in Col. 6 of the MINITAB worksheet in Figure 1.5 (previous chapter), they can be identified as being in or outside of the limits 98.82–100.82 mg:

Wt mg	100.3	99.7	100.7	97.2	98.5	97.2	97.6	99.8	99.5
	in	in	in	out	out	out	out	in	in

This shows that five of the deliveries were within the 1% tolerance limits and four were outside. This distribution could obviously be altered by the width of the limits chosen.

The above sorting job can also be carried out by MINITAB. First prepare a column C8 to receive the output and label it *Accurate*. Go to **Calc > Calculator**, and when the dialog box opens:

---

1) In the box *Store result in variable:* enter $\boxed{Accurate}$ by highlighting and selecting;

2) In the box *Expression:* enter $\boxed{\text{'Wt mg'}<100.82 \text{ And 'Wt mg'} >98.82}$; which is the shorthand code for asking the destination column to be given a 0 if the value in C7 is outside the limits, and a 1 if it is within the specified limits; note the use of the quotes to identify the variable;

3) Click on $\boxed{OK}$.

4) Observe the output in Col. 8 (Figure 1.5) as a series of 1s and zeros, where 1 = within the limits and 0 = outside.

As with the manual method, there are five 'accurate' results (those within the 1% tolerance limits) and four outside.

3

Are Those Differences Significant?

Read not to contradict and confute, nor to believe and take for granted, nor to find talk and discourse, but to weigh and consider.

Francis Bacon (1561–1626)

3.1   SIGNPOSTING FOR DIFFERENT TYPES OF DATA

In the last chapter a set of 135 weight observations was treated as one large group, without taking account of the three different pipettes or the 15 students that were involved. The object of this and the next chapter is to investigate whether there were *significant differences* between either the individual pipettes or the individual students in the average amounts of water delivered. This will require discussing the word 'significant', which is a central term in statistics and has a special meaning. But before that, it is useful to pause and consider the several types of data that the experimental biologist is likely to encounter. What follows in this chapter is suitable for only one of the four main categories of data to which statistical analysis can be applied.

3.1.1   Normally-distributed measurements

The pipetting weight is a convenient example of a very common type of quantitative observation, namely normally-distributed measurements. These are data that 1) are not restricted to whole-number values and can have as many decimal places as the measuring device allows, and 2) when a large group of them is assembled, give a symmetrical histogram whose top outline approximates to a normal distribution curve. However, instead of just judging the shape of the histogram by eye, there are objective tests (§3.4) that allow a more exact statement about the goodness-of-fit to a normal distribution.

The statistical 'toolkit' for dealing with normally-distributed measurement data is very extensive. So if data are normally distributed, it is important to know about it at an early stage. The ANOVA and *t*-test procedures depend on data being at least a reasonably close approximation to a normal distribution. Likewise when one obtains the mean and st. dev. of a set of measurements, the values are only useful if the assumption about normality is valid. For example, with a highly asymmetric distribution, the mean will not go through the top of the peak and the st. dev. will not extend to the points of inflection on the 'underlying' distribution curve. Nor will ± 1 st. dev. around the mean enclose 68% of the population. The message with measurement data is therefore to use when possible the statistical methods that are based on the normal distribution, but to be alert to departures from normality.

3.1.2   Non-normal measurements

There are some types of measurement data to which normal distribution statistics are not applicable, at least not without some adjustments. One type is where the results contain less than (<) or greater than (>) values. For example, in antibody titrations of serum, the first dilution in a titration series may be 1/10. However, if a sample contains no detectable antibody at this dilution, one can not really say that the true value is zero. All one can do is to record the reciprocal titre as <10. Likewise in a cancer treatment trial where the length of survival of the patients is recorded, the data may contain observations such as >5 years. Eventually, of course, all the patients will die but maybe not until the normal lifespan has been reached. So, for purposes of recording the results of an effective treatment, there may be many patients where the record shows >5 years. Data with either '<' or '>' signs can not be processed in the same way as the pipette weight deliveries since

MINITAB will not deliver descriptive statistics on columns of data containing such entries.

In these cases the appropriate 'measure of central tendency' is not the mean but the median. The median does not have the equivalent of a st. dev. or a SE mean as measures of variability and exactitude respectively. However it is possible to get 95% confidence intervals of the median. And it is also possible to do the equivalent of ANOVA and *t*-tests, so all is not lost. The category of statistical methods that use medians rather than means is known as *non-parametric statistics*. A *parameter* can be taken as 'a constant that occurs in the algebraic equation that defines a distribution'. For example, the equation for the normal distribution has the two parameters of mean (μ) and st. dev. (σ). So when one goes into non-parametric statistics one is avoiding making assumptions about an underlying normal (or any other) distribution. Non-parametric statistics are discussed in Chapter 5. Note that 'parameter' has crept into popular speech as a trendy word without the exact quantitative connotations as in statistics. So that one reads about 'the parameters' of selling teabags or of playing golf.

Another type of measurement data for which normal distribution methods are not appropriate are those that follow a *lognormal* distribution. This category includes many types of human and animal physiological data. For example, whereas plasma glucose concentrations in healthy people are normally distributed, plasma insulin concentrations follow a lognormal distribution. Likewise serum antibody concentrations tend to be lognormally distributed (in addition to often presenting a problem with '<' values). If the only problem is the lognormal distribution, then the conversion of the raw data to the logarithmic values is the solution. Thereafter all the data processing can be performed as for normal distribution statistics, with back conversion out of logarithms at the end. Again, this is discussed in Chapter 5.

3.1.3 Counts and proportions

These two categories of data differ from those above in being restricted to whole-number values. Examples include bacterial colony counts, numbers of insects on a leaf, and proportion of patients who were cured by a treatment. In each of these the raw data numbers can not have a decimal. So, for example, if 14/19 patients (= 73.684 21%) benefited from a treatment, the statistical analysis has to be carried out on the 14/19 and not on the 73.7%, no matter how many decimal places the calculator delivers. Such *proportion* data are discussed in Chapter 7, while the *count* data are considered in Chapter 6. For the former, the 'underlying' distribution is the binomial and for the latter, the Poisson.

The rest of this chapter is devoted to data like the pipetting weights that follow an 'underlying' normal distribution without transformation or other alteration.

3.2 ONE-WAY ANALYSIS OF VARIANCE

Books on statistics usually do not introduce Analysis of Variance until somewhere beyond Chapter 3. This is because at a theoretical level there is a lot of background to be covered and equations to be used if you really wish to understand it in depth. However, I believe it is possible to sidestep the theory for initial access, and to apply analysis of variance in an introductory fashion. There are, however, some specialist terms to be noted, and a basic idea about probability to be understood.

3.2.1 Application of MINITAB

As the initial working example, the nine pipetting weights from Chapter 1 will be taken. These are already on the worksheet in Chapter 1 (Figure 1.5) in a form suitable for analysis of variance, namely as a column of nine *Wt mg* observations (C6) with the corresponding pipette numbers (C5) alongside. The usual abbreviation for analysis of variance is ANOVA, and its purpose is to find out whether the three pipette means are significantly different from one another. The output from ANOVA is a probability statement, that states 1) whether the means are significantly different from each other, or 2) whether the amount of variation is *within the limits of random-sample fluctuations*. Both 'significant' and 'random-sampling fluctuations' are explained in the Further Notes at the end of this chapter.

From a design standpoint, the simple pipetting experiment has only one feature (what MINITAB calls a 'factor') that was imposed by the experimenter; and that is the categorization according to pipette. This qualifies the data for *one-way ANOVA*. With the larger set of 135 deliveries there was categorization by the two independent criteria ('factors') of 'pipette' and 'student'. These latter data qualify for *two-way ANOVA*.

To use MINITAB to provide one-way ANOVA of the nine observations, go to **Stat > ANOVA > One-Way** and when the dialog box appears (Figure 3.1):

1) In the *Response* box, enter *'Wt mg'*; this can either be typed in with the quotes, or preferably by highlighting *Wt mg* in the columns box, and either double-clicking, or by clicking on *Select*. Alternatively the entry can be typed in as C6 (without quotes), but this is less useful since it will appear just as 'C6' on the final printout and as an axis label on any graphs that are done;

2) In the *Factor* box enter *'Pip No'* by a similar procedure;

3) As part of the procedure it is useful to ask MINITAB for a *point plot* output of the data so as to have a pictorial display. This is obtained by clicking on the $\boxed{Graph}$ button in the *One-Way ANOVA* dialog box and, when the internal dialog box (Figure 3.1) opens, selecting *Dotplots of data*;

4) Click on $\boxed{OK}$ to close the internal box, and then on $\boxed{OK}$ again to close the main box;

5) Observe the output as the graph in Figure 3.2 and as a printout in the *Session* window. This has been copied into Table 3.1, where column numbers have been added for convenience.

Taking the graph first, it is the same as the point plot (Figure 1.9) from the first chapter, except that MINITAB has usefully added the pipette mean*s* as short bars. MINITAB is inconsistent in its terminology. If the *Graph* window is pulled down and a dotplot requested, MINITAB produces Figure 3.3, which is like Figure 3.2

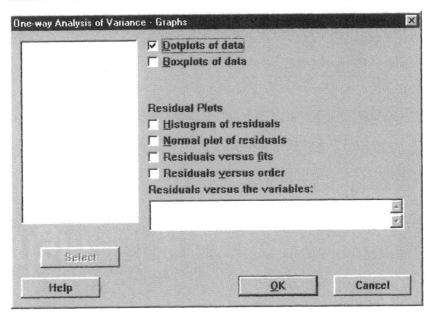

Figure 3.1 MINITAB dialog boxes for one-way analysis of variance, with entries made for analysing the results of the pipetting experiment of Chapter 1. The top screen is the main dialog box, and the lower is the box accessed by clicking on *Graphs*, with the request entered for *Dotplots of data*

Dotplots of Wt mg by Pip No
(group means are indicated by lines)

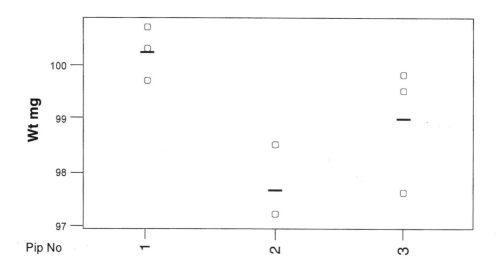

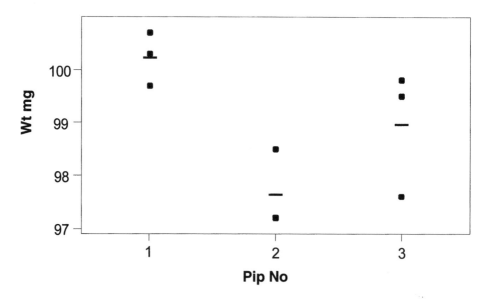

Figure 3.2   Point plots of the nine pipetting results with individual pipette means inserted.
Top, output as delivered by MINITAB; bottom, the same after editing to make the figure
suitable for publication in a journal

Table 3.1 MINITAB output for the one-way ANOVA of the nine pipetting observations of Chapter 1. Column numbers have been added for convenience of reference

Col. 1 Source	Col. 2 DF	Col. 3 SS	Col. 4 MS	Col. 5 F	Col. 6 **P**
Pip No.	2	10.142	5.071	6.79	0.029
Error	6	4.480	0.747		
Total	8	14.622			

				Individual 95% CIs for mean based on pooled st. dev.
Col. 7 Level	**Col. 8** N	**Col. 9** Mean	**Col. 10** st. dev.	**Col. 11**
				--------+---------+---------+--------
1	3	100.233	0.503	(-------*-------)
2	3	97.633	0.751	(-------*-------)
3	3	98.967	1.193	(-------*-------)
		Pooled st. dev. = 0.864		--------+---------+---------+--------
				97.5 99.0 100.5

but turned through 90° and without the mean bars. So to avoid confusion, graphs like Figure 3.2 are referred to as a point plot since asking directly for a dotplot yields Figure 3.3. This is not a major issue, but worth noting.

Setting aside the semantics, the plot in Figure 3.2 shows features that should be confirmed in the analysis of variance:

1) The mean of *Pip 1* is above the highest values of each of the two other pipettes, especially *Pip 2*;
2) The mean of *Pip 2* is well below the lowest value of *Pip 1*, and below two of the three values of *Pip 3*;
3) The mean of *Pip 3* is below the lowest value of *Pip 1* and above the highest value of *Pip 2*.

It therefore looks to the naked eye as if the three pipettes are indeed delivering 'significantly' different volumes of water, but superimposed on this is considerable scatter due to experimental inconsistencies. The virtue of ANOVA is that it wraps

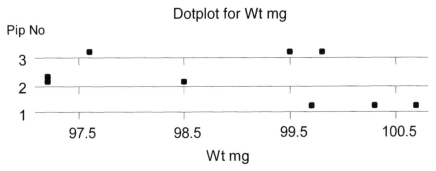

Figure 3.3 Dotplots of the nine pipetting results, displayed by pipette

up all these considerations into a single result, the *P-value*, where '*P*' stands for probability. This main result, $P = 0.029$ is boxed for prominence in Table 3.1 and is the first item to be looked for in an analysis of variance.

3.2.2 Interpretation of *P*-values

To convert the *P*-value into a verbal conclusion, the following rules are applied:

1) If *P* is greater than ($>$) 0.05, the differences in pipette means are declared to be 'not significant';
2) If *P* is less than or equal to ($\leq$) 0.05, the differences in pipette means are interpreted as being 'significant';
3) If *P* is less than or equal to ($\leq$) 0.01, the differences in pipette means are said to be 'highly significant'.

In the present example therefore, the differences are in the second category, of being 'significant'. This means that the observed differences between the pipette means can not reasonably be explained by the amount of variability in the three observations associated with each pipette, i.e. 'random-sampling fluctuations'. The next section provides some further insights into *P*-values, after which the rest of the output of the ANOVA in Table 3.1 will be considered.

3.2.3 Altered data sets

A useful way to gain further insights into *P*-values is to work with manipulated data. For example, if the pipette means are made closer together by adding 2 mg to each result from Pipette 2 and 1 mg to each result from Pipette 3, the point plots (Figure 3.4) then show a high degree of overlap. To the naked eye it would appear

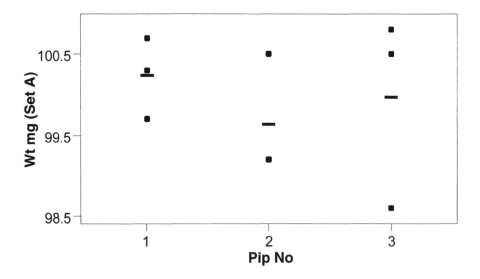

Figure 3.4 Point plots of pipetting data that were manipulated to *reduce* the between-pipette differences as compared with Figure 3.2 (see text)

Table 3.2 MINITAB output for the one-way ANOVA of the nine pipetting observations of Chapter 1, altered to be less disparate (Set A). This was done by adding 2 mg and 1 mg respectively to each value delivered by Pipettes 2 and 3 respectively. Column numbers have been added for convenience of reference

Col. 1 Source	Col. 2 DF	Col. 3 SS	Col. 4 MS	Col. 5 F	Col. 6 **P**
Pip No.	2	0.542	0.271	0.36	0.710
Error	6	4.480	0.747		
Total	8	5.022			

Individual 95% CIs for mean based on pooled st. dev.

Col. 7 Level	Col. 8 N	Col. 9 Mean	Col. 10 st. dev.	Col 11
				-------- + --------- + --------- + --------+
1	3	100.233	0.503	(------- * -------)
2	3	99.633	0.751	(------- * -------)
3	3	99.967	1.193	(------- * -------)
		Pooled st. dev. = 0.864		-------- + --------- + --------- + --------+
				99 100 101 102

that the amount of scatter in the replicates from each pipette is sufficiently large to obscure any slight differences there may be between the pipette mean*s*. This is confirmed by the ANOVA (Table 3.2) which shows a *P*-value of 0.710 that is firmly in the zone of non-significance. If the manipulations had been taken to the point of making the three means identical (by adding or subtracting suitable exact constants) the *P*-value would reach its maximum of 1.0. Thus the *higher* the *P*-value, the *lower* is the significance of the differences between the mean*s*.

Conversely, the further apart the mean values become, the lower is the *P*-value. This is shown in Figure 3.5 and Table 3.3 with a *P*-value of 0.001, reflecting highly significant differences between the pipette mean*s*. In Further Notes, FN 3.2 discusses what is meant by 'significant difference' in more detail.

3.2.4 The rest of the ANOVA output

As has been stated before, MINITAB produces abundant output, often much more than is needed for the purpose in hand. So far, the focus has been on the 'bottom line' of ANOVA, the final *P*-value and its relationship to 'significance'. The rest of the output will now be explained briefly, while there is a fuller explanation in the Further Notes section. It is most convenient to inspect Table 3.1 from Col. 11 backwards to Col. 1. Before this, it would be useful to have read the Further Notes (FN 2.1 and FN 2.2) at the end of the last chapter, about st. dev. and variance.

- *Col. 11* is like the point plot (Figure 3.2) but turned through 90° and without the individual points. It presents the mean (asterisk) of each pipette, together with the 95% confidence intervals of the mean, as curved brackets. It shows that Pipette 2 is substantially lower than Pipette 1, and with their confidence intervals

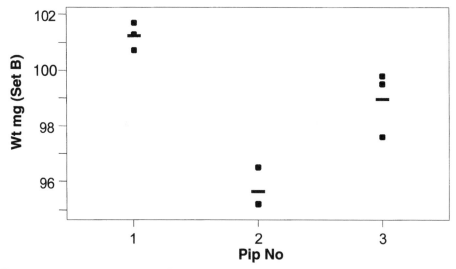

Figure 3.5 Point plots of pipetting data that were manipulated to *increase* the between-pipette differences as compared with Figure 3.2 (see text)

Table 3.3 MINITAB output for the one-way ANOVA of the nine pipetting observations of Chapter 1, altered to be more disparate (Set B). This was done by adding 1 mg to each value from Pipette 1 and subtracting 2 mg from each value delivered by Pipette 2. Column numbers have been added for convenience of reference.

Col. 1 Source	Col. 2 DF	Col. 3 SS	Col. 4 MS	Col. 5 F	Col. 6 **P**
Pip No.	2	49.609	23.804	31.88	**0.001**
Error	6	4.480	0.747		
Total	8	52.089			

				Individual 95% CIs for mean based on pooled st. dev.
Col. 7 Level	Col. 8 N	Col. 9 Mean	Col. 10 st. dev.	Col. 11
				-------- + --------- + --------- + --------+
1	3	101.233	0.503	(---- * ----)
2	3	95.633	0.751	(---- * ----)
3	3	98.967	1.193	(---- * ----)
		Pooled st. dev. = 0.864		-------- + --------- + --------- + --------+
				95 97.5 100 102.5

not overlapping. Pipettes 1 and 3, and 2 and 3 overlap to some extent but with the confidence intervals of each pair only just touching the other's mean. Overall this type of picture is consistent with the differences between the pipette means being *significant*, but not *highly significant*, as reflected in the *P*-value of 0.029.

- *Col. 10* contains the st. devs of the data from each pipette. It is important for the validity of ANOVA that the st. devs of the individual groups of data should not be too different from each other. There is an objective test for this, the *Bartlett test for homogeneity of variances*, that is available on MINITAB. It should be activated if there is suspicion that the st. devs may be too different from each other. With such small ($N = 3$) data sets, a difference of twofold or threefold in st. devs is not a matter for concern. Note that ANOVA uses a *pooled st. dev.* This is not the simple mean of the three individual st. devs but is obtained by averaging the variances and taking the square root. Thus the 0.864 for pooled st. dev. is obtained from:

$$\text{pooled st. dev.} = \sqrt{[(\text{st. dev. } 1)^2 + (\text{st. dev. } 2)^2 + (\text{st. dev. } 3)^2]/3} \qquad \text{Eq. 3.1}$$
$$= \sqrt{[(0.503^2 + 0.751^2 + 1.193^2)/3]}$$
$$= \sqrt{[2.24026/3]}$$
$$= 0.864$$

- *Col. 9* contains the mean for each pipette and Col. 8 the number of observations from each pipette.
- *Col. 7* labelled 'Level' refers to the three pipettes as the 'factor' in ANOVA.
- *Col. 6* has as its sole entry the probability P from the ANOVA and is discussed further in FN 3.2.
- *Col. 5* contains the *variance ratio* (F) obtained from Col. 4 by dividing the first entry in the *MS* column by the second, i.e. $5.071/0.747 = 6.79$.
- *Col. 4* labelled '*MS*' contains two *mean squares* (*MS*), otherwise known as *variances*. Variance is defined as the st. dev. squared. So the pooled st. dev. = 0.864 in Col. 10 is the square root of the '*error MS*' = 0.747 in Col. 4. The terminology of ANOVA is quite confusing for non-statisticians, because:
  1) The purpose of ANOVA is to analyse mean values for their possible differences;
  2) The term *mean square* is used in place of variance although the *MS* values are *variances*, and the '*F*' in Col. 5 is a *variance ratio*;
  3) The actual analytical process of ANOVA involves separating the *sums of squares* (*SS*) and *degrees of freedom* (*DF*) in Cols 2 and 3 of the ANOVA table.
- *Col. 3* The formulae for calculating these quantities are not given here. But note that the 'error' SS is the same 4.480 in all three of Tables 3.1, 3.2 and 3.3. This is because the error SS reflects the extent of variation of the replicate weight deliveries from each pipette, which was not altered by the process of adding or subtracting constants from the pipette means. A feature of MS values is that the 'Pip No' MS and the 'error' MS add up to give the 'total' MS.
- *Col. 2* DF are degrees of freedom and are discussed in FN 2.1. However it should be noted here that the *Pip No DF* = 2 and the *error DF* = 6 add up to give the *Total DF* = 8. All three of the ANOVA tables are identical in this respect.
- Finally *Col. 1* labelled 'Source' refers to the *source of the variation*. The 9 *Wt mg* observations have two independent sources of variation, the variation due to possible differences between the pipettes (*Pip No*), and the variation that emerged as experimental 'error' in making the replicate observations from each pipette. The word 'error' does not mean 'mistake' but is a technical term in statistics for the sum-total of the unidentifiable influences in experimental work that cause replicate observations not to be identical. So the total variation is made up of the two components *between-pipette* variation and *within-pipette* (or *error*) variation.

Working now from Col. 1 forwards, the procedure of ANOVA separates the total variation into its two components of pipette variation and error variation. This is achieved by splitting up the *total DF* into *pipette DF* and *error DF*. Likewise the *total SS* is split into *pipette SS* and *error SS*. Then the MS entries are obtained by dividing each SS by its DF, e.g. in the first line 10.142/2 = 5.071 and likewise 4.480/6 = 0.747. In Col. 5 the pipette MS is divided by the error MS to give F= 6.79. In the days before MINITAB, this so-called *found* value of F would be compared with the value of F looked up in statistical tables, and the decision about significance based on the comparison.

The above is a very abbreviated account that attempts to strike a balance between a proper theoretical presentation and allowing use of MINITAB in an essentially empirical manner. For ANOVA to be used with validity the criteria to be met are that:

1) The data should have been collected by a procedure involving randomization;
2) The 'underlying population' should be normally distributed, or approximately so;
3) The st. devs of the individual groups should not be too different and if in doubt checked by the Bartlett test (§ 6.7.3).

3.3 WHICH PIPETTE IS DIFFERENT FROM WHICH?

The output probability of $P = 0.029$ from the ANOVA that was applied to the nine pipetting results states that the three pipettes are significantly different from each

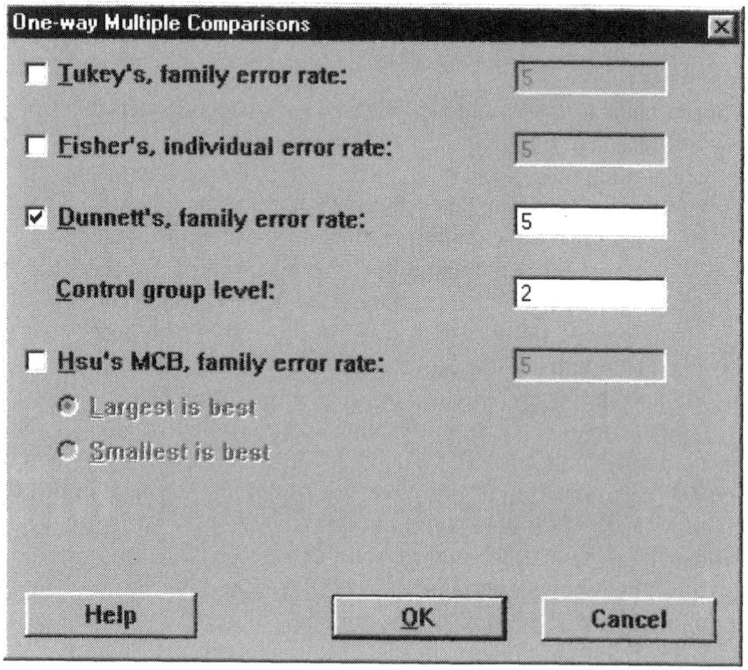

Figure 3.6 Dialog box for Dunnett and Tukey tests, accessed by clicking on *Comparisons* in the main analysis of variance box (Figure 3.1, top), with the request entered for *Dunnett* and specifying Pipette No. 2 as the *Control group*

other, but does not specify which is different from which. To explore this question requires additional tests of which the two most useful are the Dunnett and the Tukey tests.

Dunnett is useful if one of the experimental groups is a baseline, or standard, of some kind. For example, in a physiological experiment, there may be a baseline group of tissues to which no treatment is given and which serves as a control for other groups, some of which are expected to give larger physiological responses. Dunnett compares each of the treatment groups with baseline and determines which are significantly different from it. Dunnett is not really applicable in the pipetting experiment since none of the pipettes was identified beforehand as a standard with which the other two were to be compared. However, to demonstrate the mechanics of Dunnett, let us take Pipette No. 2, which had the lowest average delivery, as the notional baseline for the comparisons.

Tukey, on the other hand, compares all the pipettes with each other and indicates which is different from which. Both Tukey and Dunnett use the conventional $P = 0.05$ (or 5%) as the conventional boundary between significant and not significant. Both tests are available in an internal dialog box of the one-way ANOVA.

3.3.1 Dunnett test

Return to §3.2.1 and repeat the initial steps of the one-way ANOVA, i.e. go to **Stat > ANOVA > One-Way** and when the box appears:

---

1) In the *Response* space, enter '*Wt mg*';
2) In the *Factor* space enter '*Pip No*';
3) Press the [*Comparisons*] button and, when the internal dialog box opens,
4) Click on *Dunnett*;
5) In the space opposite *Control group level*, enter [2] (for pipette 2, which is being taken as the baseline);
6) Click on [*OK*] to close the internal box and again on [*OK*] to close the main ANOVA box.

The Session window should now show the Dunnett output as in Table 3.4.

The output from Dunnett is the *difference in means* of the baseline group (here, taken as Pipette 2) and of each compared group (Pipettes 1 and 3). The scale at the bottom right shows the differences in the means and, to each of these differences, is attached the 95% confidence intervals indicated by the curved brackets. A significant difference between groups (i.e. pipettes, referred to as *Levels*) is indicated when the span of the CI values *does not enclose zero* on the difference scale. This happens with Pipette 1 ('Level' 1), in its comparison with Pipette 2, where the lower CI is 0.58 and the upper CI is 4.6. This may be seen on the graph, or read more accurately from the printout to the left. In contrast, the difference of means of Pipettes 3 and 2 has CI values (–0.68 to 3.35) that straddle the zero mark, indicating that Pipette No. 3 is not significantly different from No. 2.

Table 3.4   MINITAB output for the Dunnett test applied to the three groups of three pipetting results of Chapter 1. The important results are in bold

Dunnett's intervals for treatment mean minus control mean
Family error rate = 0.0500
Individual error rate = 0.0287
Critical value = 2.86
Control = level (2) of Pip No

Level	Lower	Center	Upper				
				-------- + --------- + --------- + --------+			
1	**0.5802**	**2.6000**	**4.6198**		(------------- * -------------)		
3	**-0.6865**	**1.3333**	**3.3532**	(------------- * -------------)			
				-------- + --------- + --------- + --------+			
				0.0	1.5	3.0	4.5

A practical point in setting up an experiment in which Dunnett is eventually to be used is that the pre-identified baseline group should contain more observations than any of the other groups. This is because of its use as a reference point for multiple comparisons and the desirability of having its position fixed with greater exactitude. The rule is that if you have 'K' groups to be compared with a baseline, then the number of observations in the baseline group should be $\sqrt{K}$ times as large as any of the groups with which it is to be compared. So if there are four treatment groups and one baseline group, the baseline should have twice ($\sqrt{4} = 2$) as many replicate observations as any of the treatment groups. Therefore a preferred design would be to have two replicate baseline groups, each with the same number of observations as each treatment group, and randomized along with them.

3.3.2   Tukey test

This test is done in a similar way to Dunnett, except that there is no baseline group to be specified, since each group is compared with all the others. With the three pipettes, the output from Tukey is as in Table 3.5. The layout of the results is different from Dunnett but the interpretation is similar, in that significant differences are those where the confidence intervals do not enclose a zero difference in the means. This happens only with the Pipette 1 versus Pipette 2 comparison. The two other comparisons, P1 versus P3 and P2 versus P3, both have CIs that enclose zero and are therefore not significant. In applying these (and other statistical) tests it is useful to have the point plot in mind, since any serious discrepancy between visual appearance and statistical output should be carefully scrutinized to see if a procedural error has been made.

3.4   TESTING FOR NORMALITY

The treatment of the pipetting data as samples from an 'underlying' normal distribution has provided a consistent thread through this chapter, and ANOVA depended on this assumption being valid. Support was provided by the histogram of the $N = 135$ data set in the last chapter and also the calculations where the number of observations that lay within ± 1 st. dev. and ± 1.96 st. dev. of the mean was very close to expectations. MINITAB does in addition provide specific tests for

Table 3.5 MINITAB output for the Tukey test applied to the three groups of three pipetting results of Chapter 1. The important results are in bold

Tukey's pairwise comparisons
Family error rate = 0.0500
Individual error rate = 0.0220
Critical value = 4.34
Intervals for (column level mean) – (row level mean)

	95% Confidence intervals of the difference in the means for the comparison:	
	Pipette 1	**Pipette 2**
Versus Pipette 2	**0.4348 to 4.7652**	
Versus Pipette 3	**–0.8985 to 3.4318**	**–3.4985 to 0.8318**

normality (i.e. conformity to a normal distribution) which will now be explored. It should be pointed out that with a small data set of only $N = 9$ observations, only very serious departures from normality will be detected. Larger data sets such as the $N = 135$ observations are much more sensitive to revealing departures from normality.

The essence of these procedures is to convert the bell-shaped normal distribution curve into a straight line and then see how closely the experimental points fit it. This inspection can be done initially by eye. There is also a statistical test for goodness-of-fit, with the usual probability value of 0.05 (or 5%) being taken as the arbitrary boundary between significant and non-significant departures from normality. Note however that it is not possible to *prove* normality but only to detect significant departures from it, or to say that there were *no* significant departures from normality. There is a similarity here with the chemical purity of, for example, a sample of sodium chloride. Thus one may 1) detect impurities if they are present in sufficient amount to give a response in the analytical method used, or 2) be able to say that there are no impurities at a particular level of detection. But one can not say that the sample is absolutely pure down to the last atom of sodium or chlorine. The normality test is similar in allowing the possible conclusions 1) that *departures* from normality are detectable or 2) that they are not detectable in the data supplied.

To perform the normality test on the nine pipetting deliveries of Chapter 1, open the MINITAB worksheet from Chapter 1 that has the nine weighings of pipette deliveries, then go to **Stat > Basic Statistics > Normality Test** and, when the dialog box (Figure 3.7) appears:

1) In the *Variable* space, enter '*Wt mg*';
2) Of the three tests of normality offered, in the first instance select the *Ryan-Joiner*;
3) Press the OK button.

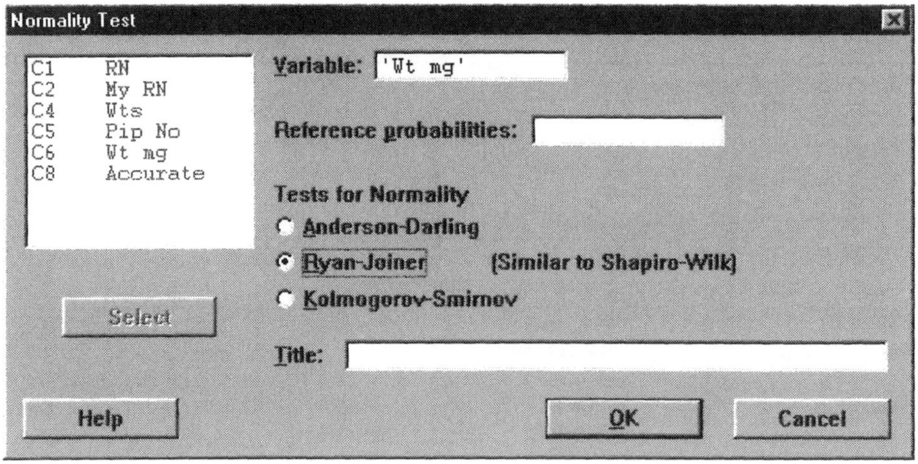

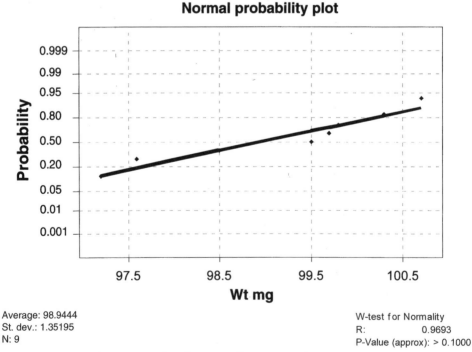

Figure 3.7   Dialog box for normality test on the Pipette *Wt mg* data, with Ryan-Joiner as the specific test requested

The output (Figure 3.8) is a plot of *Wt mg* against *Probability*. The fitted straight line is that given by a bell-shaped normal distribution curve that has gone through two stages of algebraic processing. First it was converted to an S-shaped cumulative-probability distribution curve, which in turn was made linear by conversion to the spread-out scale on the probability axis. Effectively the probability

Normal probability plot

Average: 98.9444
St. dev.: 1.35195
N: 9

W-test for Normality
R: 0.9693
P-Value (approx): > 0.1000

Figure 3.8   MINITAB output for the normality test on the nine pipetting results, with Ryan-Joiner statistics

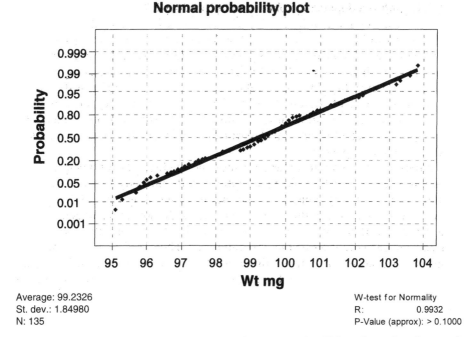

Average: 99.2326
St. dev.: 1.84980
N: 135

W-test for Normality
R: 0.9932
P-Value (approx): > 0.1000

Figure 3.9 MINITAB output for the normality test on the 135 student pipetting results, with Ryan-Joiner statistics

scale is collecting the area under the normal curve from left to right, from a very small left tail which reaches down to 0.001, until close to 1.0 has been reached in the right tail.

Note that 0.5 on the probability scale corresponds to a weight of about 99 mg, the mean of the data, and the centre of the normal distribution. Note also that all except the two extreme points fall within the probability range of about 0.2 and 0.8, corresponding to a span of 0.6 (or 60%), or around ± 1 st. dev. on either side of the mean. Thus the graph is to be interpreted as showing experimental points not too distantly scattered about the theoretical normal line produced from data with mean 99.94 and st. dev. 1.35. The main feature to look out for is any tendency of the experimental points to show systematic curvature away from the theoretical straight line. None is visible here.

The statistical test for goodness-of-fit is the *P*-value of >0.1 which, being above the conventional 0.05, is not significant. Thus this outcome does not prove that the sample of nine results is normally distributed; rather that with a sample of this small size, any departures from normality were insufficient to be detected.

If instead of the Ryan-Joiner test for normality, the other two tests offered by MINITAB are selected, the output is exactly the same graph but the probability values are different, although still not significant. Anderson-Darling gives *P* = 0.252, and Kolmogorov-Smirnov has *P* >0.15.

The larger data set with 135 pipetting results provides a much more stringent test of normality. As happened above, the three normality tests give the same graph (Figure 3.9), but with different *P*-values. Anderson-Darling gives *P* = 0.021, which indicates a *significant* departure from normality, as does Kolmogorov-Smirnov with $P = 0.017$. However Ryan-Joiner yields $P > 0.1$ which is not significant. At first sight this seems to be the sort of result that gives statistics a bad reputation and allows the comment 'You can prove anything with statistics...'. However, the three tests operate on different assumptions and are sensitive to different types of departure from normality. The chemical analogy would be different tests for, say, potassium in a sample of sodium chloride. One test might fail to detect impurity while two others might be sufficiently sensitive to find definite positive traces. The fact that two tests detect significant departures from normality suggests that such departures are indeed present.

The experimental points in Figure 3.9 show no evidence of following a *curved* distribution, such as will be seen in Chapter 5 with antibody titres. Rather, the departure from normality consists of the graph having long runs of 10 or more points on one or other side of the straight line. This is the sort of result generated by a data set where there are two or more normal distributions that are slightly displaced from each other along the horizontal axis. There is already evidence from the nine-result experiment that the three pipettes delivered significantly different volumes. This is confirmed in the larger data set of 135 results where each pipette contributed 45 observations. So although the histograms of the 135 results only hinted at the three overlapping peaks, because of their closeness together, the normality plot and two of the tests provide evidence for their existence. The reason that Ryan-Joiner did not indicate significance is because it is sensitive for picking up curvature, which these data do not show. On the other hand, Anderson-Darling and Kolmogorov-Smirnov are sensitive to excessive flattening of the normal distribution curve that occurs when the underlying distribution has separate peaks, as here.

When the 45 results of each pipette were tested for normality as three separate groups, none revealed a significant departure from normality by any of the three tests. Therefore there is continued justification for using normal distribution statistics for the analysis of these data, even though the 135 results as a group are slightly non-normal for the reason uncovered.

This section is concluded by presenting the results (Figure 3.10) of a normality test on the 5000 genuine normally distributed values (Figure 2.8 of the previous Chapter) produced by MINITAB from a mean = 99.2 and st. dev. = 1.85, the same as the 135 pipetting results. Here the points, except for a few at the ends, all lie exactly on the theoretical straight line, and there is no significant departure from normality by any of the three statistical tests.

3.5 SUMMARY

This chapter started with a section on the different types of experimental data – measurements, counts and proportions that require different statistical toolkits. Logically, this material should be at the beginning of the book so that the signpost to normal distribution statistics would have already been passed before starting the first pipetting experiment. Be that as it may, the main substance of this chapter has been the one-way analysis of variance, and the Dunnett and Tukey tests that follow from it. There is a presentational conflict between going straight into MINITAB to

Normal probability plot

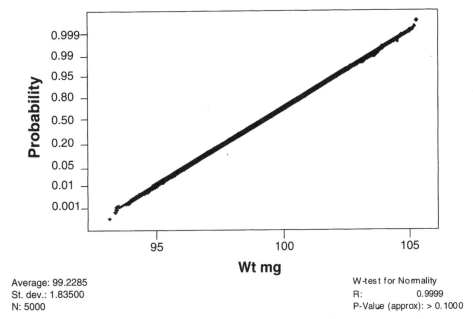

Average: 99.2285
St. dev.: 1.83500
N: 5000

W-test for Normality
R: 0.9999
P-Value (approx): > 0.1000

Figure 3.10 MINITAB output for the normality test on 5000 random samples from a normal distribution with the mean and st. dev. the same as for the 135 pipetting result, and with Ryan-Joiner statistics

perform ANOVA, which can be done quite readily, but in a 'black box' fashion and, on the other hand taking the reader through the theory of ANOVA and spending substantial time with formulae and pocket calculators. The main text is deficient in some of the important background aspects that the Further Notes are intended to remedy. Therefore these additions should be seen as a full part of the main narrative and of equal importance, but segregated for separate consideration. Thus the approach has been 'to do first and understand afterwards', and to blend the graphical–pictorial with the statistical–numerical. The chapter ended with the objective testing for normality because of the underpinning assumption that the data fitted a normal distribution.

FURTHER NOTES

FN 3.1   Random-sampling fluctuations

In statisticians' eyes, the nine-result pipetting experiment of Chapter 1 produced a *sample* from a theoretical 'underlying' *population* of data that has no existence except as an idea. MINITAB produced the mean and the st. dev. of this sample, as estimates of the population values, i.e. the parameters μ and σ respectively. Further investigation by ANOVA showed, however, that the situation was more complicated in that the pipettes appeared to be significantly different. A more accurate theoretical model would be of three overlapping populations with means of μ_1, μ_2

and μ_3 for the three, slightly different, pipettes but with a shared standard deviation of σ, reflecting the 'error' variation introduced by the human operator and the way each pipette was used. MINITAB was not able to say that the three pipettes were *definitely* different from each other, but only to make a probability statement about it. The *P*-value of 0.029 delivered by ANOVA, being below the conventional $P = 0.05$ boundary, showed a 'significant' difference between the pipettes.

So what does 'significant' mean? To gain some insights, let us generate some random sets of $N = 9$ observations from a normal distribution with mean = 98.944 (the mean weight of the nine pipette deliveries) and st. dev. = 0.864, the error st. dev. in the ANOVA. The error variation reflects the variation associated with the replicates from each pipette after allowing for the between-pipette differences. Let us generate 20 such sets of random data, since the idea of 'significant' reflects an event that occurs on average no more often than once in every 20 random samples. Go to **Calc > Random Data > Normal** and, when the dialog box appears (Figure 3.11):

---

1) In the *Generate* space, enter $\boxed{9}$;
2) *Store in column(s)*: $\boxed{\text{C1-C20}}$;
3) *Mean:* $\boxed{98.944}$;
4) *Standard deviation:* $\boxed{0.864}$;
5) Press the $\boxed{OK}$ button.

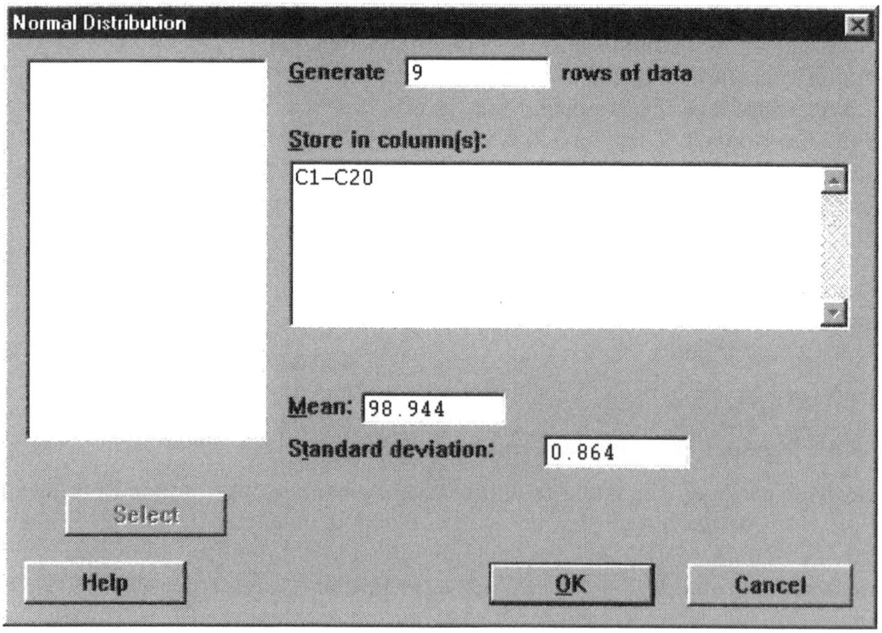

Figure 3.11   Dialog box for generating nine rows and 20 columns of random *Wt mg* data from a normal distribution with mean 98.944 and st. dev. 0.864

Table 3.6 Twenty sets (A1–A20) of $N = 9$ randomly distributed values from a normal distribution with mean 98.944 and st. dev. 0.864. The entries have been shortened to one decimal place. ANOVA of the A4 set (bold), when analysed as three groups of three observations, gave a P-value of 0.053. Figure 3.14 is a point plot of Set A4

A1	A2	A3	A4	A5	A6	A7	A8	A9	A10	A11	A12	A13	A14	A15	A16	A17	A18	A19	A20	Pip. No.
97.5	100.0	99.2	**97.8**	98.5	97.4	98.2	98.4	99.4	99.1	99.4	99.6	98.5	98.2	100.0	100.1	99.6	99.3	97.9	97.9	1
98.6	97.3	98.4	**99.0**	98.7	99.4	97.9	98.7	97.8	97.8	97.9	97.9	100.6	99.9	98.7	99.9	98.9	99.2	98.2	98.8	1
100.4	98.2	100.1	**98.6**	97.1	97.6	98.4	99.9	98.9	98.2	99.3	98.7	98.9	99.8	97.9	97.7	99.6	99.1	98.4	99.2	1
99.6	98.1	100.0	**98.8**	98.6	98.9	99.5	97.6	99.3	98.8	100.6	97.5	99.2	100.6	100.1	98.4	100.2	97.8	100.	99.9	2
99.1	98.5	98.9	**99.6**	99.3	99.1	97.5	99.0	97.9	100.5	99.6	98.6	97.4	99.3	97.2	98.6	99.9	98.9	99.9	97.5	2
98.5	99.2	99.1	**99.5**	98.6	99.2	98.5	99.6	98.4	98.7	99.5	99.1	96.7	98.8	99.3	98.0	97.8	98.0	98.5	99.3	2
97.9	99.4	98.7	**98.2**	99.0	100.2	98.4	99.8	98.6	98.4	100.2	99.1	97.9	98.0	98.0	99.3	99.3	98.7	99.2	99.4	3
99.8	98.6	98.4	**98.5**	97.7	97.9	100.9	100.7	98.2	99.8	98.6	99.5	100.1	99.3	99.2	99.4	99.9	98.4	99.0	98.5	3
98.1	99.4	97.7	**97.7**	99.8	99.3	99.3	98.2	97.3	98.6	99.3	98.1	98.4	97.6	99.4	97.5	98.5	97.3	99.0	99.8	3

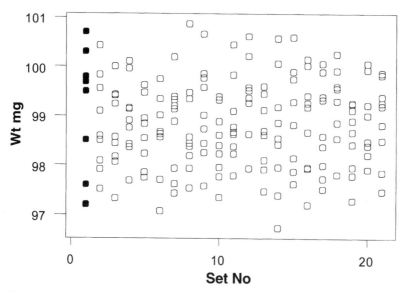

Figure 3.12   Point plots of the nine pipetting results of Chapter 1 (solid squares) and 20 sets of nine random samples from a normal distribution with the same mean and error st. dev. (open squares)

Table 3.6 is a record of 20 nine-weight sets of random data from a normal distribution with mean = 98.944 and st. dev. 0.864. The set A4 (in bold) is discussed below because it is the one set in the sample of 20 where an apparently significant 'difference in pipettes' has arisen purely through random-sampling fluctuations.

To the point plots of these 20 sets of random data (Figure 3.12), the nine 'real' pipetting observations are added as Set No. 1 in bold. The plots show that the extent of scatter in the nine 'real' observations is not particularly different from those of the 20 generated sets of nine random observations. The 20 generated data sets thus illustrate 'random-sampling fluctuations' in samples of $N = 9$ observations from a normal distribution with the stated parameters.

A different picture emerges when the data in each of the 20 random sets are treated, not as nine ungrouped observations, but as replicates of three from each of three 'pipettes'. With the 'real' data, the association of particular weights with particular pipettes emerged as actual experimental results. With the 20 random sets, each was displayed on the MINITAB worksheet in a column of nine random weights. Therefore when a column of notional pipette numbers was placed vertically alongside as 1, 1, 1, 2, 2, 2, 3, 3, 3, each notional pipette was allocated an already randomized value. So it would only be by chance that particular notional pipettes would be given weights that were clustered in a way that looked like genuine differences between pipettes.

Figure 3.13 is a plot of the three mean weights for each 'notional pipette' in the 21 data sets. This is revealing. It shows that although some of the 20 random sets had pipette means that were quite widely scattered, none was as widely scattered as the real data, Set No. 1, displayed in bold.

One-way ANOVA was then performed on each of the 20 fictitious data sets to

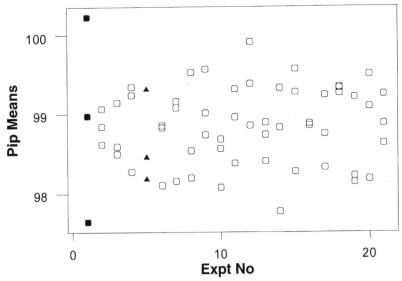

Figure 3.13 Point plots of the individual pipette means of the nine pipetting results of Chapter 1 (solid squares) and of corresponding group means of 20 sets of nine random samples from a normal distribution with the same mean and error st. dev. (open squares, and solid triangles for Set A4)

see if any would give a 'significant' result. The expectation is that in random sampling such as that done here, there should be an *average* of one set in every 20 that gives a P-value of 0.05 or less. This one-in-20 figure refers to a long run of, say, hundreds of samplings, not 20 as performed here. So it would be a matter of luck (chance) if the particular set of 20 contained one such 'significant' result. In fact, Set No. 4 came close to it with an $F = 4.97$ and $P = 0.053$. A point plot of this set in Figure 3.14 shows data that the eye might well judge as indicating a significant difference between 'pipettes'. But of course this result arose purely by chance as an uncommon event and as a reflection of random-sampling fluctuations. Therefore when making a judgement about 'real' experimental data, the extent of between-group differences has to be judged against the background variation as seen here. The $P = 0.029$ given by the 'real' data *could* have arisen by a random-sampling fluctuation, the probability being about once in every $1/0.029 = 35$ data sets in a long run of random sampling.

Note that Set 4 does not appear to have especially widely scattered means in Figure 3.14, but this is compensated by a very small error variance (0.214). This gave a relatively large F-value (4.97) and correspondingly small probability ($P = 0.053$) that was just marginally out of the *significant* zone.

FN 3.2 Null hypothesis

The previous section highlighted the dilemma that emerges when trying to make a judgement about group differences in real data, against the background of random-sampling fluctuations. The final output of the ANOVA with the real pipette results was the P-value of 0.029 and refers to the variance ratio, $F = 6.79$. This F-value is

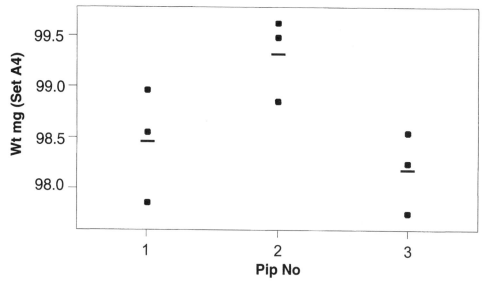

Figure 3.14   Point plots of the nine results of random Set A4 (solid triangles in Figure 3.13) in random groups of three, and their means

the ratio of the *between-pipette variance* to the *error variance* and expresses in a single number the extent to which the pipette means are separated from each other, in relation to the amount of experimental error. Here is the dilemma: 1) a *P*-value of 0.029 could be expected to arise once in every 35 'experiments' purely by the vagaries of random-sampling fluctuations; 2) under discussion is one 'real-life' experiment that gave a *P*-value of 0.029. So is there a *real* difference between pipettes, or is the result a 1 in 35 random chance? The answer is that it is not possible to be *sure*, but it is possible to make a probability statement.

This brings in the null hypothesis (NH), a standard statistical procedure, which states in this example that 'any apparent differences in pipette means *is not real*, but has arisen through random-sampling fluctuations'. The $P = 0.029$ is the probability of getting $F = 6.79$ *if* this NH is correct. Purely by convention, *P*-values of 0.05 and 0.01 have come into usage as boundaries for accepting or rejecting the NH. If *P* is greater than 0.05 the NH is accepted and the apparent between-group differences declared to be 'not-significant'. If *P* is less than or equal to 0.05, the NH is rejected and the differences declared to be 'significant'; if *P* is less than or equal to 0.01, the differences are described as 'highly significant'. Thus the present result of $P = 0.029$ leads to the rejection of the NH and the conclusion that the three pipettes delivered average volumes that are significantly different at the $P = 0.029$ level.

An individual experiment with a *P*-value only slightly below the 0.05 boundary (as this one here) might be described as 'showing significant differences', but in practice not too much reliance should be placed on it. However, if the experiment is repeated several times and continues to deliver significant differences, then the conclusion becomes correspondingly more convincing. As will be seen in the next chapter, when ANOVA is applied to the 135 pipetting results, where there are 45 replicate observations from each pipette, the *P*-value associated with between-pipette differences is less than 0.001. Even this result does not exclude the remote chance that the $P = 0.001$ reflects the one chance in 1000 that the result could have arisen by a random-sampling fluctuation.

It is therefore important to appreciate that 'significant' is used with a special and precise meaning in statistics that is rooted in probability theory. In the present case, the conclusion that the different pipettes were delivering 'significantly different' volumes of fluid might not actually be 'significant' in the day-to-day work of the laboratory. Because unless the work was of very high precision, the minor differences between the pipettes might not actually be 'significant' in the everyday usage of the word.

4

More About Measurement Differences

If your experiment needs statistics, you ought to have done a better experiment.

Lord Rutherford (1871–1937)

4.1 INTERNAL ANATOMY OF THE 135 PIPETTING RESULTS

The 135 pipetting results have already been used in the previous chapters but without inspecting their 'internal anatomy', i.e. the role of the three different pipettes and the 15 students in determining the overall structure of the data. This chapter therefore starts with analysis of the 'pipette' and 'student' factors, first pictorially and then by *two-way ANOVA*. The pipetting data have no special merit in themselves. Their main purpose is to serve as a general example of some of the statistical procedures that can be applied to a fairly large group of measurement data with two independent variables, or 'factors' (*pipette* and *student*).

The later part of the chapter is devoted to the special case of comparing group means where there are only two groups. Although this can also be done by *one-way ANOVA*, the usual procedure is *Student's t-test*.

Throughout this chapter, the assumption is made that the data are all *measurements*, as distinct from *counts*, and that there is no strong evidence of departures from normality.

4.1.1 Point plots

To obtain point plots of the 135 results, open the MINITAB worksheet where they are recorded (Table 2.2) and then use the procedure from §1.7, except applying it to the larger data set and taking the *Jitter* option straightaway.

Go to **Graph > Plot**, and when the dialog box (Figure 4.1) opens:

1) At $\boxed{Graph\ variables}$ and opposite $\boxed{Graph\ 1}$, enter '*Wt mg*' in the $\boxed{Y\text{-}column}$ and '*Pip No*' in the $\boxed{X\text{-}column}$;
2) Check that the $\boxed{Data\ display}$ box contains opposite $\boxed{Item\ 1}$ the default entries of *Symbol* under $\boxed{Display}$, and *Graph* under $\boxed{For\ each}$;
3) Click on $\boxed{Options}$, and when the sub-box opens,
4) Click on the tick space adjacent to $\boxed{Add\ Jitter\ to\ Direction}$;
5) Alter the default option to 0.05 $\boxed{X}$ and change the $\boxed{Y}$ to 0.0001;
6) Click on $\boxed{OK}$ to close the sub-box and again on the $\boxed{OK}$ in the main box.

The output in Figure 4.2, where there are 45 observations from each pipette, does not show such a clear-cut difference between pipettes as the much simpler Figure 1.7 representing the data of a single student. There is perhaps a tendency for Pipette No. 2 to be slightly lower than Nos 1 and 3, but the differences are not convincing. Nevertheless it will be seen in the ANOVA (§4.3) that the differences between pipettes are statistically significant.

The same point plot procedure was applied to exploring possible differences between students, by inserting '*Stud No*' instead of '*Pip No*' in the *Plot* menu. Here *Jitter* was initially invoked but then omitted as the point plot (Figure 4.3) looked better without it. This plot reveals that there were major differences between students, both in the mean weights and in the extent of scatter.

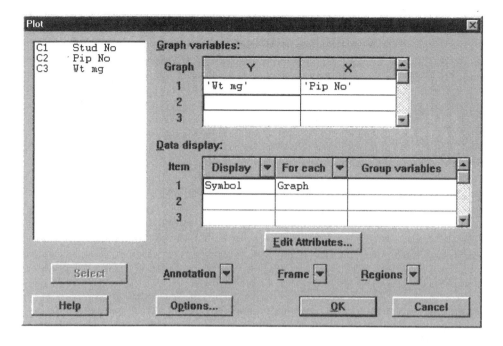

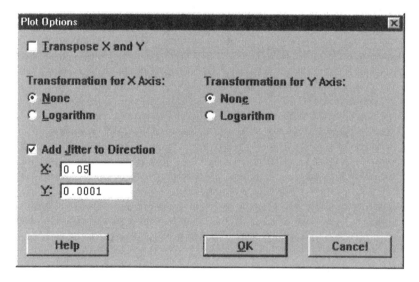

Figure 4.1 MINITAB dialog boxes for point plots, by *Pip No*, of the 135 pipetting results. Top, main box; bottom, sub-dialog box for adding *Jitter*

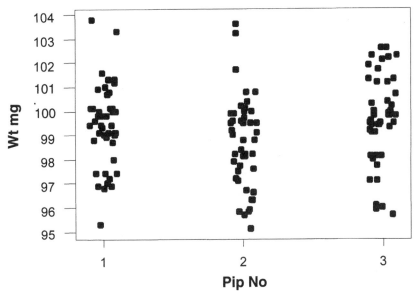

Figure 4.2 Point plots, with *Jitter*, of the 45 *Wt mg* values delivered by each of three pipettes

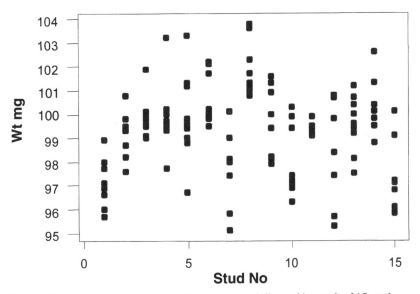

Figure 4.3 Point plots of the nine *Wt mg* values delivered by each of 15 students

4.1.2 Boxplots (default)

Another very useful type of pictorial output from MINITAB is the *boxplot*, which is available with various options and could have been introduced with the introductory experiment in Chapter 1. It is also known as a *box-and-whiskers* plot. To apply it, go to **Graph > Boxplot**, and when the dialog box (Figure 4. 4) opens:

1) Start at the top space $\boxed{Graph\ variables:\ Y\ (measurement)\ vs\ X\ (category)}$

2) Opposite $\boxed{Graph\ 1}$, enter '*Wt mg*' in the $\boxed{Y\text{-}column}$ and '*Pip No*' in the $\boxed{X\text{-}column}$;

3) In $\boxed{Data\ display}$ note that the default settings under $\boxed{Display}$ are *IQRange Box* and *Outlier S>>*; also the two entries of *Graph* under $\boxed{For\ each}$;

4 Click on $\boxed{OK}$.

The output in Figure 4.5 (top) contains the boxplots for the 45 observations from each of the three pipettes. The central horizontal line in each box is the median and the horizontal boundaries of the box above and below it are the *1st* and *3rd* *quartiles* (see §1.8.2). Therefore the height of the box, between these quartiles, represents the central 50% of the data. The position of the median, whether central or towards the upper or lower edge, is an indication of the symmetry of the distribution. Here, all three medians are close to central and the distributions are therefore symmetrical. The vertical lines are the *whiskers* and are a further measure of the amount of scatter of the data in each group. With Pipette 1, there are also *outliers*, represented with asterisks. These are observations sufficiently detached from the rest of the data as to be regarded as discontinuous from them.

The boxplot thus deals in *non-parametric statistics* as there is no assumption of an 'underlying' normal distribution and therefore no representation of mean or st. dev. The median is used as the *measure of central tendency* in each group of data, and the extent of scatter is depicted as the height of the box, the length of the whiskers and the outliers. As shown below, the 95% confidence intervals of the median can be added to each boxplot to give the bottom diagram in Figure 4.5.

Figure 4.6 is the default boxplot for the *Student* factor, or variable, in the data. This was obtained in the same way as the *Pipette* factor, but with replacement of '*Pip No*' by '*Stud No*' in the procedure above. As with the point plots, it strongly indicates that there were considerable differences between individual students in the median weights of water delivered and in the degree of scatter. Student 11, for example, had an extremely consistent set of results, whereas No. 8 while not the most scattered, had excessively high values. Note in Figure 4.6 that the *Edit* menu was used to add extra thickness to the main box outlines of students 1–10, whereas the diagrams for students 11–15 are unedited from how MINITAB delivered them.

4.1.3 Boxplots (options)

The boxplot menu offered by MINITAB contains several useful options of which just a few will be explored. Figure 4.5 (bottom), with the pipettes, contains boxes nested within each other. The outer box in bold lines is the *interquartile range*, as before, while the inner box with fine lines is the *95% confidence intervals (CI) of*

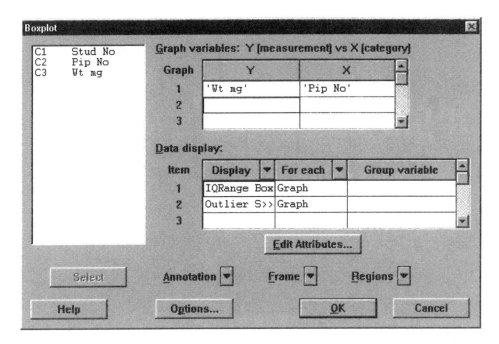

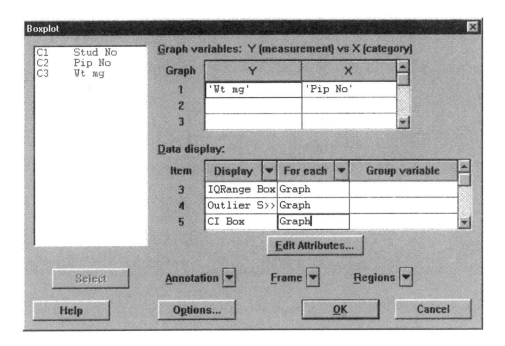

Figure 4.4 Dialog box for boxplots of the 45 *Wt mg* values from each of three pipettes. Top, main box; bottom, sub-dialog box with 95% CIs requested

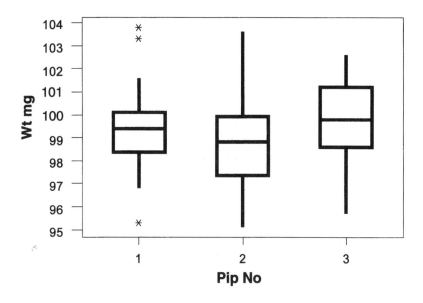

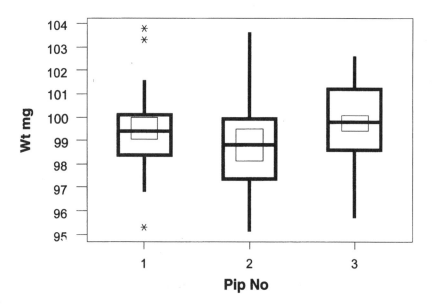

Figure 4.5 Boxplots of the *Wt mg* values from each of three pipettes. Top, default output; bottom, with 95% confidence intervals of the medians added (fine lines)

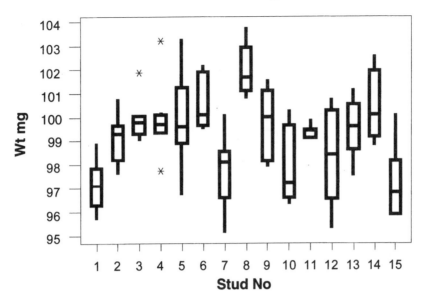

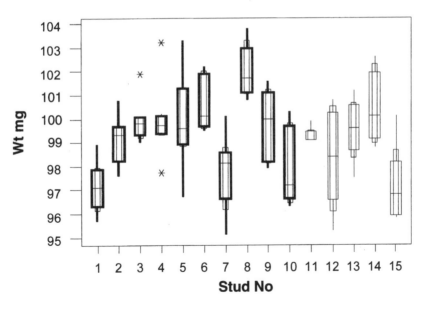

Figure 4.6 Boxplots of the *Wt mg* values from each of 15 students. Top, default output; bottom, with 95% confidence intervals of the medians added (fine lines). Extra artwork was done on students 1–10 in the bottom diagram (see text)

the median. This is obtained by the following alteration at step 3 above (see Figure 4.4, bottom):

---

3) In $\boxed{Data\ display}$ note the default settings under $\boxed{Display}$ as *IQ Range Box* and *Outlier S>>*. Also note the two entries of *Graph* under $\boxed{For\ each}$; pull down the $\boxed{Display}$ window and select the top item $\boxed{CI}$ to insert as Item 3; opposite it, and under $\boxed{For\ each}$, type in *Graph*; then proceed as previously. This gives the bottom diagrams in Figure 4.5.

An aspect for close scrutiny with the CIs is whether their extrapolated levels horizontally enclose or exclude the medians of adjacent boxes. Enclosure is interpreted as indicating lack of significant differences between the groups, while exclusion, if it is large enough, indicates significant differences. Inspection of the pipette boxplots (it helps to have a ruler to hold against the page) reveals Pipette 3 as having a median that is well above the upper CI of Pipette 2, and vice versa the median of Pipette 2 is below the lower CI of Pipette 3. These are strong suggestions of a significant difference between these two pipettes. The other two comparisons, Pipette 1 vs Pipette 2 and Pipette 1 vs Pipette 3 show some overlapping of medians and CIs and the differences are not so significant.

Note that the CI box may be enclosed within the interquartile range box or may extend beyond it. The determining factor is the number of observations. With each pipette, there are 45 observations. These make the CIs relatively narrow and therefore within the confines of the interquartile box. With the student boxplots however, the CIs extend beyond the boxes (Figure 4.6, bottom) because the number of observations from each student is only 9.

Boxplots are sometimes presented (Figure 4.7) with the *X*- and *Y*-axes transposed.This can be done by clicking on the $\boxed{Options}$ button of the main boxplot dialog box and selecting $\boxed{Transpose\ X\ and\ Y}$. This internal dialog box is then closed and the procedure completed as before. All in all, boxplots provide a useful visual impression of the 'internal anatomy' of data sets that have moderate to large numbers of observations and one or more 'factors' for classification.

4.2 SUMMARY STATISTICS

4.2.1 Descriptive statistics, by factor

The descriptive statistics may be obtained as in §1.8, but entering the data from the worksheet (Table 2.2) with the 135 pipetting results. Two lots of these descriptive statistics are worth asking for: tabulated by *pipette* and separately by *student*. Table 4.1 presents the descriptive statistics, tabulated by pipette. The means and medians of each group are very close to each other, confirming the symmetry seen in the boxplots. The st. devs show little variation between the pipettes, as would be expected.

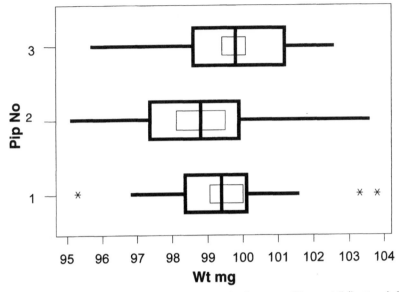

Figure 4.7 Boxplots of the *Wt mg* values from each pipette, as Figure 4.5 (bottom), but with the *X*- and *Y*-axes transposed

Table 4.1 MINITAB output of descriptive statistics for the 135 pipetting deliveries, tabulated by pipette. Column numbers have been added for convenience of discussion

Col. 1	Col. 2	Col. 3	Col. 4	Col. 5	Col. 6	Col. 7	Col. 8	Col. 9	Col. 10	Col. 11
Variable	N	Mean	Median	Tr Mean	St. Dev.	SE Mean	Min.	Max.	Q1	Q3
Wt mg 1	45	99.362	99.400	99.320	1.707	0.255	95.30	103.80	98.35	100.10
2	45	98.684	98.800	98.615	1.897	0.283	95.10	103.60	97.350	99.90
3	45	99.651	99.800	99.695	1.845	0.275	95.70	102.600	98.60	101.20

The tabulation by *student* was calculated by a different procedure so as to cut down the amount of output from MINITAB and only get the mean and the st. dev. To do this as shown in Figure 4.8, go to **Statistics > Basic Statistics > Store Descriptive Statistics**, and when the dialog box opens:

1) In Variables select '*Wt mg*';
2) In By Variables (optional): , enter '*Stud No*';
3) Click on Statistics and, when the sub-box opens, note that the default state has only three options selected: *Mean, Standard deviation* and *N nonmissing*; deactivate the last of these (bottom picture in Figure 4.8);
4) Click on OK to close the internal box, and again on OK in the main box.

Unlike what happens with *Display Descriptive Statistics*, the output from *Store Descriptive Statistics* is not delivered to the *Session* window but appears in the first two empty columns in the worksheet, to which MINITAB gives the labels *Mean 1* and *St. Dev. 1*. These may be highlighted, copied on to the clipboard and pasted on

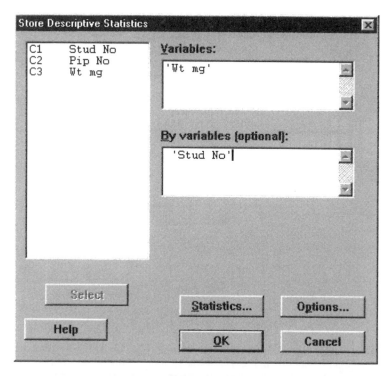

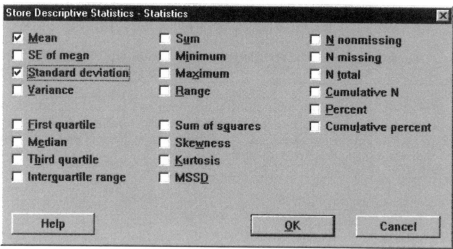

Figure 4.8　Dialog boxes (main and sub-box) for *Store Descriptive Statistics*, with entries to obtain only the mean and st. dev. of the *Wt mg* values for each student

to a new Word document (Table 4.2) with the student number alongside. Here it will be seen that the student means vary from 97.1 up to 102.0, while the st. devs go from 0.277 up to 2.06, a sevenfold range. Analysis of variance will determine whether the student means differ significantly from each other, while the Bartlett test (§ 6.7.3) can be used to explore differences in st. devs.

Table 4.2 MINITAB output of the individual student means and st. devs in the 135 pipetting deliveries (rounded to three or four significant figures)

Student No	Mean	St Dev
1	97.1	0.992
2	99.1	0.984
3	99.9	0.848
4	99.9	1.440
5	99.9	1.86
6	100.6	1.11
7	97.7	1.52
8	102.0	1.10
9	99.7	1.41
10	97.8	1.57
11	99.3	0.277
12	98.4	2.06
13	99.6	1.19
14	100.4	1.45
15	97.1	1.53

4.2.2 A feast of descriptive statistics

This section concludes with an illustration of what happens if MINITAB is asked to produce descriptive statistics with 'all the trimmings' (Figure 4.10). Actually this figure for Pipette 3, is just one-third of the output since the other two pipettes are treated similarly but are omitted here to save space. All three can be viewed on the screen simultaneously with the *Tiling* option in the *Window* window of MINITAB. To produce Figure 4.10, go to **Stat > Basic Statistics > Display Descriptive Statistics**, and when the dialog box appears (Figure 4.9):

1) Enter '*Wt mg*' in $\boxed{Variables}$;
2) Click on the small square to the left of $\boxed{By\ variable}$ so that a $\boxed{✔}$ appears;
3) Enter '*Pip No*' (with quotes) to the right of $\boxed{By\ variable}$;
4) Click on $\boxed{Graphs}$ and, when the internal box (Figure 4.9, bottom) opens, tick $\boxed{✔}$ for all the options offered, i.e.
   - ✔ *Histogram of data*
   - ✔ *Histogram of data, with normal curve*
   - ✔ *Dotplot of data*
   - ✔ *Boxplot of data*
   - ✔ *Graphical summary*
5 Click on $\boxed{OK}$ to close the internal box and on $\boxed{OK}$ again to close the main box.

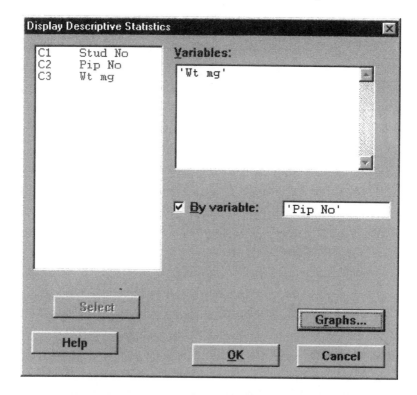

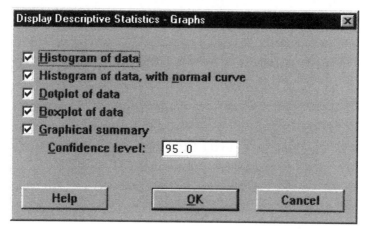

Figure 4.9   Dialog boxes for obtaining the full descriptive statistics of the *Wt mg* values from each pipette, of which the output for Pipette No. 3 is presented in Figure 4.10

The output in Figure 4.10 is probably more than will normally be wanted as a statistical summary of a single set of data. However, items to note are that the Anderson-Darling normality test gave a non-significant result ($P = 0.08$), meaning that the 45 deliveries from the individual pipette did not differ significantly from a normal distribution. The skewness and kurtosis refer to departures from normality

Descriptive Statistics

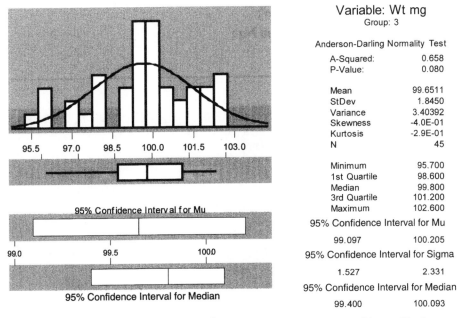

Variable: Wt mg
Group: 3

Anderson-Darling Normality Test

A-Squared:	0.658
P-Value:	0.080
Mean	99.6511
StDev	1.8450
Variance	3.40392
Skewness	-4.0E-01
Kurtosis	-2.9E-01
N	45
Minimum	95.700
1st Quartile	98.600
Median	99.800
3rd Quartile	101.200
Maximum	102.600

95% Confidence Interval for Mu

99.097	100.205

95% Confidence Interval for Sigma

1.527	2.331

95% Confidence Interval for Median

99.400	100.093

95% Confidence Interval for Mu

99.0   99.5   100.0

95% Confidence Interval for Median

Figure 4.10   Full descriptive statistics of the 45 *Wt mg* values from Pipette No. 3

that are discussed in MINITAB *User's Guide* 2, p. 1.6. The confidence intervals for the 'underlying' population mean Mu (μ) are wider than those for the median.

4.3   TWO-WAY ANALYSIS OF VARIANCE

4.3.1   Application of MINITAB

Two-way analysis of variance may be applied to the 135 pipetting results to answer four questions:

1) Are there significant differences in the mean deliveries from the three pipettes?
2) Are there significant differences in the mean deliveries from the 15 students?
3) Is there significant *interaction* between pipettes and students?
4) What is the value for error st. dev., i.e. the average amount of scatter within each of the 45 triplicate values of *Wt mg* produced by nine students with three pipettes?

For the analysis to be valid, the data must be random samples from one or more 'underlying' normal distributions, and there should not be significant differences in the error st. dev., i.e. *within* the 45 sets of triplicate observations generated by 15 students each using pipettes. To conduct the ANOVA, the data should be set out as in Table 2.2, where each of the 135 *Wt mg* values has, in adjacent columns, a number indicating the pipette and student with which it is associated.

For MINITAB to provide two-way ANOVA of the nine observations, go to **Stat > ANOVA > Two-way** and when the dialog box (Figure 4.11) appears:

1) In Response , enter 'Wt mg';
2) In Row factor enter 'Pip No';
3) In Column factor enter 'Stud No';
4) Click on OK to close.

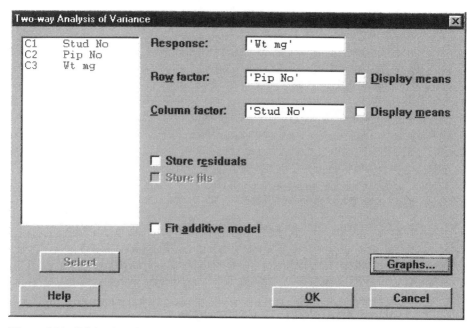

Figure 4.11 Dialog box for two-way analysis of variance, with entries for *Pip No* and *Stud No* as the independent variables, and *Wt mg* as the dependent variable

Table 4.3 presents the results of the two-way ANOVA. In Col. 6 there are three probability (*P*) values of 0.000, in other words all below 0.001 and therefore *very highly* significant. Thus the null hypotheses for *pipettes*, for *students* and for the third term, *interaction*, are all rejected, and the differences can not be ascribed to 'random-sampling fluctuations' (see FN 3.1). However, the analysis so far does not say which individual pipettes are different from which, and likewise which students are different from each other. The term *interaction* is described in §4.3.2 below.

Two-way ANOVA does not offer the equivalent of the Dunnett and Tukey tests for further comparisons. However, some indications of significant differences between individual pipettes and students can be found in the *confidence limit* diagrams (Table 4.3, middle and bottom). The rule to be applied (to a first approximation) is that if the mean is well separated from the CIs of the compared group then the two groups are significantly different at the *P* = 0.05 level, as is implied by 95% CIs. Thus Student No. 8 is very different from most of the others. Likewise No.1 is significantly different from No. 2.

Table 4.3 MINITAB output from the two-way ANOVA, by pipette and by student, of the 135 pipette deliveries of *Wt mg*. Column numbers have been added for convenience of discussion

Col. 1 Source	Col. 2 DF	Col. 3 SS	Col. 4 MS	Col. 5 F	Col. 6 P
Pip No	2	22.16	11.08	9.68	0.000
Stud No	14	237.77	16.98	14.83	0.000
Interaction	28	95.53	3.41	2.98	0.000
Error	90	103.05	1.15		
Total	134	458.52			

```
                         Individual 95% CI for pipettes
Pip No       Mean      -+---------+---------+---------+---------+
  1          99.36                   (-------*-------)
  2          98.68       (-------*-------)
  3          99.65                      (-------*-------)
                        -+---------+---------+---------+---------+
                        98.40   98.80   99.20   99.60   100.00
```

```
                         Individual 95% CI for students
Stud No      Mean      --------+--------+--------+--------+---
  1          97.11     (---*---)
  2          99.07         (---*----)
  3          99.88             (---*----)
  4          99.90             (---*----)
  5          99.90             (---*----)
  6          100.57             (----*---)
  7          97.74     (-------*----)
  8          102.00                      (---*----)
  9          99.71       (---*----)
 10          97.86         (---*----)
 11          99.28            (---*----)
 12          98.37       (----*---)
 13          99.57          (---*----)
 14          100.43            (----*---)
 15          97.11     (---*---)
                      --------+--------+--------+--------+---
                      97.6      99.2    100.80  102.40
```

4.3.2 Meaning of interaction

The pictorial representation of the significant *interaction* term in the ANOVA is provided in Figure 4.12, which is a point plot of the three pipette means for each student. Pipette 1 is represented by solid squares, No. 2 by open squares and No. 3 by solid triangles. If there were *no* interaction, then the points for the three pipettes would be in the same vertical order in each student column, i.e. invariably Pipette X at the top, Pipette Y next and Pipette Z the lowest. But instead of that regular arrangement, Pipette 1 is top, or top-equal, in only seven columns, while eight columns have Pipette 3 as top, or top-equal, and Pipette 2 is top or top-equal three times. In the lowest, or lowest-equal position, Pipette 1 occurs once, Pipette 2 occurs ten times and Pipette 3 is found three times.

Interaction thus expresses *the extent of sequence irregularity* of the individual

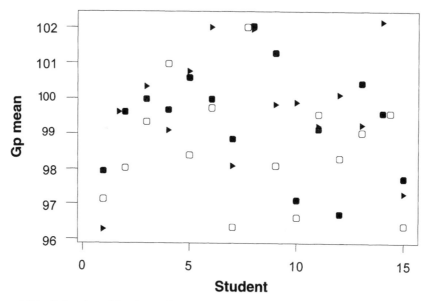

Figure 4.12   Point plot of the three pipette means from each of 15 students, to illustrate the term *interaction* (see text)

pipettes in the average volumes delivered from them by different students. This irregularity, while statistically significant, is not enough to obscure the genuine differences between pipettes and there is reasonable consistency in Pipette 2 delivering the smallest average volume in the hands of most students. This is also reflected in the pipette boxplots (Figure 4.5).

Expressed another way, there would be no significant interaction if *all* students had without exception found Pipette 3 to deliver the largest average volume, with Pipette 1 next and Pipette 2 the smallest. Numerically *zero* interaction would require not only the pipettes to be in the same order with each student, but for the *relative* volumes delivered to be the same. So that if some students found the middle-delivering pipette (No. 1) to be closer to the top-delivering pipette (No. 3), while others found it closest to the lowest-delivering pipette (No. 2), there would be some interaction but it would probably not be statistically significant.

4.4   TWO-WAY ANOVA ON FEWER DATA

4.4.1   A two-student data set

There are various other aspects of two-way ANOVA that are best explored with a data set smaller than the 135 pipetting results. Therefore for these further insights, the results of just two students, Nos 7 and 13, have been selected from Tables 2.1 and 2.2. These are to be treated as an entirely new set of 18 observations. They can usefully be set up either as a new MINITAB project or as a new worksheet within the same project as the 135 pipetting results. These 'new' sets of results with their descriptive statistics are recorded in Table 4.4, while the point plots are given in Figure 4.13.

Table 4.4 Layout of MINITAB worksheet with the pipetting data from two students (Nos 7 and 13) as a small data set for further exploration of two-way ANOVA

Col. 1 Stud No	Col. 2 Pip No	Col. 3 Wt mg	Col. 4 Spacer	Col. 5 Pipette	Col. Pip Mean	Col. 7 Student	Col. 8 St Mean	Col. 9 Wt mgX
7	1	99.0		1	99.6333	7	97.7444	99.0
7	1	100.1		2	97.6833	13	99.5667	100.1
7	1	97.4		3	98.6500			97.4
7	2	98.1						98.1
7	2	95.8						95.8
7	2	95.1						*
7	3	98.1						98.1
7	3	98.0						98.0
7	3	98.1						98.1
13	1	100.7						100.7
13	1	99.4						99.4
13	1	101.2						101.2
13	2	100.0						100.0
13	2	99.6						99.6
13	2	97.5						97.5
13	3	100.4						100.4
13	3	98.1						98.1
13	3	99.2						99.2

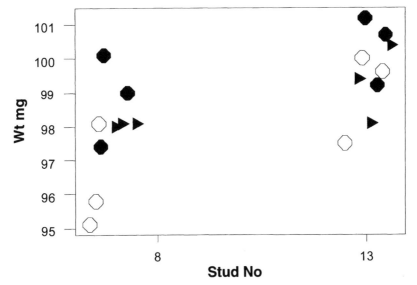

Figure 4.13 Point plot of the nine *Wt mg* values from two selected students, with the deliveries from each of three different pipettes represented by different types of point

To provide better horizontal separation of the points, the *Jitter* option on the *X*-coordinate was increased to 0.1 while the *Y-Jitter* was entered as 0.0001. Some extra artwork was added to provide different types of graph point for the different pipettes. This was done on-screen, after the point plot had been obtained, by clicking on the right-hand button on the mouse. This opened the *Edit* toolkit for making the changes. Each point was identified by consulting the list of means in Col. 6 of Table 4.4 and was then highlighted and adjusted from the menu of graph points offered by the toolkit.

Visual inspection of the point plot suggests that there are definite differences between the means of the two students and also in the means of the three pipettes which are all shifted upwards by Student 13. It also looks as if there might be some degree of interaction, probably not statistically significant, since the pipette means (as roughly interpolated by eye) of Student 7 are more spaced out than those of Student 13.

4.4.2 Another choice for two-way ANOVA

Two-way ANOVA can be carried out by a method different from that above, with the so-called *general linear model (GLM)*. This is a very high-powered statistical procedure that is frequently omitted from even quite advanced textbooks because the formulae and calculations would be too complex and tedious to do with a pocket calculator. Yet there it is in the MINITAB toolkit just waiting to be used. It is a procedure with great versatility, being capable of handling many complex experimental designs, including those with missing observations, that ordinary two-way ANOVA can not cope with. As will be seen, it gives a slightly different and more useful output than the standard two-way ANOVA when applied to the data set in hand.

To enter GLM, go to **Stat > ANOVA > General Linear Model** and, when the dialog box (Figure 4.14) appears:

---

1) In $\boxed{Responses}$ enter 'Wt mg';
2) In $\boxed{Model}$ enter 'Stud No'|'Pip No' with the two variables separated by the vertical line-bracket, and without spacing;
3) Click on $\boxed{OK}$ to close the box.

The output from GLM, in the Session window, is as copied into the top half (**a**) of Table 4.5. The 'bottom line', as with other ANOVAs is the column of *P*-values that are highlighted in bold. They show that there was a highly significant ($P = 0.007$) difference between the two students, and a significant ($P = 0.043$) difference between the three pipettes. These differences refer, of course, to the means of each student and of each pipette. The ANOVA table has an extra column in it for adjusted sum of square, which in this example is the same as the sequential sum of the square in the previous column.

The GLM ANOVA table also shows the interaction term with a *P*-value = 0.526, which is completely non-significant. Thus the output of the ANOVA is consistent with the provisional conclusions that can be derived from visual inspection of the point plots in Figure 4.13.

When the same data were put through two-way ANOVA as in §4.3.1 above, the output was as recorded in the lower section (**b**) of Table 4.5. Here there is no interaction term. Why? The reason is that if the interaction *P*-value in a two-way

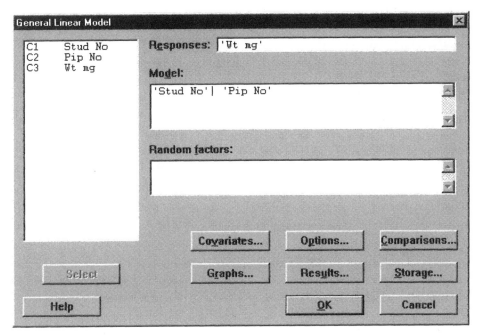

Figure 4.14 Dialog box for analysis of variance using the general linear model, with entries for *Wt mg* to be analysed by *Student* and *Pipette* as the 'factors'

ANOVA turns out to be non-significant, it is common practice to incorporate the interaction *DF* and *SS* into the error *DF* and *SS* respectively. Inspection of the GLM and two-way ANOVA outputs show that this has been done. The 2 *DF* and the *SS* = 1.847 associated with *Stud No*Pip No* (the interaction term) in the GLM table have been added in to the error *DF* and *SS* of the lower table. The effect of this is to give a larger divisor when calculating the *F*-value from the ratio of Stud No/Error *MS* and Pip No/Error *MS*. The larger *F*-values then correspond to lower *P*-values, so that the effect of the whole manoeuvre is to give the *Student* and *Pipette* terms slightly more 'weight' than they otherwise would have. Some statisticians do not approve of MINITAB taking this extra step without being so requested. They would prefer to see the actual non-significant *P*-value for inter-action displayed in the ANOVA output and then make any further slight adjustments on a discretionary basis.

4.4.3 Missing values

A big advantage of GLM over the standard two-way ANOVA is its ability to cope with a data set that is deficient in one or more observations. Since it is not un-common for observations to be 'lost' during an experiment, this is a very useful feature. As an exercise, make a duplicate of the *Wt mg* column and delete one of the 18 observations, as follows: highlight Col. 3 by clicking on the mouse and dragging downwards to capture all the entries. Go to the **Edit** menu and select **Copy Cells**. Move the cursor over to the empty Col. 9 on Table 4.3 and paste in the

Table 4.5   MINITAB output of two-way ANOVA of the two-student pipetting results by a) general linear model ANOVA and b) two-way ANOVA

a) General Linear Model

Factor	Type	Levels	Values
Stud No	fixed	2	7 13
Pip No	fixed	3	1 2 3

Analysis of variance for Wt mg, using adjusted SS for tests

Source	DF	Seq SS	Adj SS	Adj MS	F	P
Stud No	1	14.942	14.942	14.942	10.80	**0.007**
Pip No	2	11.408	11.408	5.704	4.12	**0.043**
Stud No*Pip No	2	1.874	1.874	0.937	0.68	**0.526**
Error	12	16.600	16.600	1.383		
Total	17	44.824				

b) Two-way ANOVA

Analysis of variance for Wt mg

Source	DF	SS	MS	F	P
Stud No	1	14.94	14.94	11.32	**0.005**
Pip No	2	11.41	5.70	4.32	**0.035**
Error	14	18.47	1.32		
Total	17	44.82			

```
                            Individual 95% CI
Stud No     Mean      +---------+---------+---------+---------+
7           97.74     (-------*-------)
13          99.57                  (--------*--------)
                      +-----------+-----------+-----------+-----------+
                      97.00    98.00       99.00      100.00     101.00


                            Individual 95% CI
Pip No      Mean        +---------+---------+---------+---------
1           99.63                    (---------*---------)
2           97.68       (---------*---------)
3           98.65          (---------*---------)
                        +---------+---------+---------+---------
                        97.00    98.00   99.00    100.00
```

data, labelling the column *Wt mgX* to distinguish it from *Wt mg*. Then delete observation No. 6 and try the two-way ANOVA as in §4.3.1 but with *Wt mgX* as the response. The outcome is an error message saying that the unbalanced design is not analysable. If the same is now done with GLM, the analysis is delivered but with considerably altered *P*-values. Thus the loss of just one result out of the 18 causes the *P*-value for *Student* to increase to 0.015, which is still significant but no longer highly significant, while the *Pipette* *F*-value is 0.095 which is now no longer significant. The interaction term at $P = 0.486$ was scarcely affected. Thus while GLM can cope with a missing observation, the discriminatory power of the ANOVA to detect differences in means is considerably diminished.

4.5 STUDENT'S T-TEST

4.5.1 Special case of two groups

If there are only two groups of normally-distributed measurements whose means are to be compared for significance of the difference, the best procedure is Student's *t*-test. It was devised by an English chemist, W. S. Gossett, while working at the Guinness Brewery in Dublin in the early years of the 20th century. For reasons that would probably seem quaint today, he used the pseudonym *Student* rather than his own name for authorship of the paper, which was published in a statistics journal in 1908. It is probably the most widely used of the statistical *tests of significance*, which include the *F*-test, the *chi-square* test and the Mann-Whitney test. The *t*-test effectively does the same thing as one-way ANOVA but is restricted to comparing the means of two groups of measurement data. Thus ANOVA has much more general application, to many groups of data, and can cope with more than one variable. However in the special case of comparing the means of two groups, ANOVA and the *t*-test arrive at the same conclusions.

4.5.2 Two groups of pipetting data

This section continues with the pipetting data as exemplifying a wide variety of normally-distributed measurements. The working example is the triplicate *Wt mg* values from Pipettes 1 and 2, as used by Student No. 7 in Cols 2 and 3 of Table 4.4. These data were copied into Cols 1 and 2 of a new MINITAB worksheet as shown in Table 4.6, which also has additional material. Figure 4.11 is a point plot, which suggests that the two groups are possibly different but, since there is some overlap, it is difficult to know whether the difference will be significant. This is an instance

Table 4.6 MINITAB worksheet for *t*-test experiments (see text)

Col. 1 Pip No	Col. 2 Wt mg	Col. 3 Pipette	Col. 4 Weight	Col. 5 Pip 1&3	Col. 6 Pip 1&4	Col. 7 Pip 1&5	Col. 8 Pip 1&6
1	99.0	1	99.0	99.0	99.0	99.0	99.0
1	100.1	1	100.1	100.1	100.1	100.1	100.1
1	97.4	1	97.4	97.4	97.4	97.4	97.4
2	98.1	2	98.1	97.1	96.1	95.1	99.1
2	95.8	2	95.8	94.8	93.8	92.8	96.8
2	95.1	2	95.1	94.1	93.1	92.1	96.1
		3	97.1				
		3	94.8				
		3	94.1				
		4	96.1				
		4	93.8				
		4	93.1				
		5	95.1				
		5	92.8				
		5	92.1				
		6	99.1				
		6	96.8				
		6	96.1				

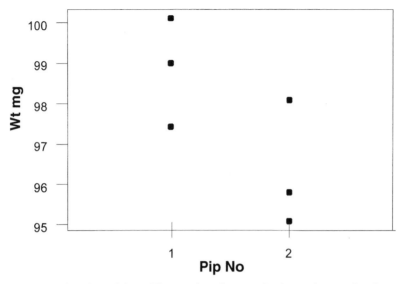

Figure 4.15   Point plots of three *Wt mg* values from each of two pipettes, for the *t*-test

where using the *t*-test provides objective criteria for reaching a decision where the naked eye is in doubt.

To perform the *t*-test, go to **Stat > Basic Statistics > 2-Sample t** and, when the dialog box (Figure 4.16) opens:

1) Select ● *Samples in one column* ;
2) Enter '*Wt mg*' opposite *Samples* ;
3) Enter '*Pip No*' opposite *Subscripts* ;
4) Leave the default option *Not equal* in the *Alternative* box;
5) Enter a ✔ at *Assume equal variances* ;
6) Click on *OK* to close the box.

Table 4.7   MINITAB output for the *t*-test on the weights of water delivered by the two pipettes from Table 4.6. Line numbers have been added to assist discussion

Line 1	Two Sample T-Test and Confidence Interval
Line 2	Two sample T for Wt mg

	Pip No	N	Mean	St. dev.	SE Mean
Line 3	Pip No	N	Mean	St. dev.	SE Mean
Line 4	1	3	98.83	1.36	0.78
Line 5	2	3	96.33	1.57	0.91

Line 6	95% CI for mu (1) – mu (2): (–0.83, 5.83)
Line 7	T-Test mu (1) = mu (2) (vs not =): T = 2.09 P = 0.11 DF = 4
Line 8	Both use Pooled St. Dev. = 1.47

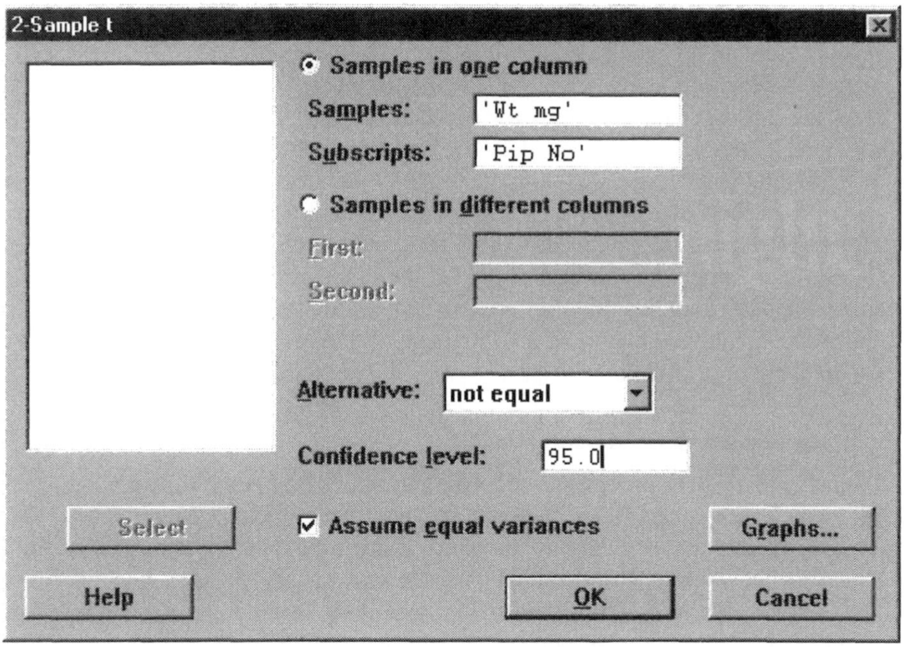

Figure 4.16   Dialog box for two-sample *t*-test applied to the *Wt mg* values from each of two pipettes

The output in the Session window is as in Table 4.7. Lines 1–5 identify that the *t*-test is being carried out on the variable *'Wt mg'* and then summarize the basic statistics of the two groups of data. They are both of size $N = 3$, they have means that differ by 2.5 mg, and they have very similar st. dev*s* and SE means. The similarity of the st. devs, which was apparent from the point plot, justified step 5 in the previous paragraph when defining the details of the *t*-test for MINITAB.

The rest of the output is given in very condensed language that needs some explanation. Line 6 introduces the assumption that the means ($\bar{X}_1$) and ($\bar{X}_2$) of the two pipettes are valid estimates of the 'underlying' population means μ_1 and μ_2, stated as mu(1) and mu(2), which may or may not be different. If they are *not* different, then the 95% confidence intervals (CI) of the difference will straddle zero. This is the case here, where the 95% CIs stretch from –0.83 to +5.83, i.e. straddling 0.0. So Line 6 says that the difference in mean weights is not significant at the $P = 5\%$ or 0.05 level.

Line 7 contains four separate pieces of information in very condensed format:

1) It sets up a null hypothesis which states that the two means differ by no more than can be explained by random-sampling fluctuations;
2) It then presents the value of the *t-statistic* $t = 2.09$;
3) At the end of the line there is DF = 4. The four degrees of freedom come from there being two groups of three observations each, and each group having 'lost' one DF through having had its mean calculated. Thus if the *t*-test was being carried out on a group of $N = 4$ and another group of $N = 6$ observations, the DF would be $4 + 6 - 2 = 8$.

4) Without MINITAB, the interpretation of $t = 2.09$ would have to be looked up in a *t-table* at the entry point of 4 DF. However MINITAB does all this automatically and provides the final probability value of $P = 0.11$. This is higher than the conventional $P = 0.05$ for significance, and therefore the difference in mean weight of water delivered by the two pipettes is *not significant*. The $P = 0.11$ (or one chance in nine) says that with the amount of scatter of points from each pipette (Figure 4.15), in relation to the difference in means, there is a chance of one in nine of obtaining that result purely through a random-sampling fluctuation. Therefore it does not qualify for being considered 'real' on the data available. This conclusion is determined by the actual data analysed. Thus the difference could become 'real' (i.e. $P \leq 0.05$) if N were larger, the st. dev. smaller or the difference in means bigger.

4.5.3   Agreement with one-way ANOVA

It is instructive to demonstrate directly the agreement between the *t*-test and one-way ANOVA. Using the procedure already described (§3.2.1), enter '*Wt mg*' as a *Response* and '*Pip No*' as a *Factor*, and show that the MINITAB output is as Table 4.8. See that the *P*-value is the same as for the *t*-test (with an extra decimal place) and the error st. dev. = 1.47 is likewise identical. The CI diagram in the lower right of the table is instructive in showing the two pipettes ('Levels') with their *95% CIs* as overlapping with each other's mean.

4.5.4   Manipulated data

Further insights into the *t*-test can be obtained from analysis of manipulated data in which the means of the two groups being analysed are made further apart or closer together. This has been done in Cols 5 to 8 of Table 4.6 where there are four such manipulated data sets. The first three lines of each of them contain the *Wt mg* figures of Pipette 1, entered by the *Copy* and the *Paste* functions in the *Edit* window.

Table 4.8   MINITAB output of one-way ANOVA on the six observations from Pipettes 1 and 2, as subjected to the *t*-test in Table 4.7

One-way analysis of variance
Analysis of variance for *Wt mg*

Source	DF	SS	MS	F	P
Pip No	1	9.38	9.38	4.35	**0.105**
Error	4	8.61	2.15		
Total	5	17.99			

Individual 95% CIs for Mean
based on pooled st. dev.

Level	N	Mean	St. dev.	
				+---------+---------+---------+-----
1	3	98.83	1.36	(----------*----------)
2	3	96.33	1.57	(----------*----------)
				+---------+---------+---------+-----
				94.0 96.0 98.0 100.0

Pooled st. dev. = 1.47

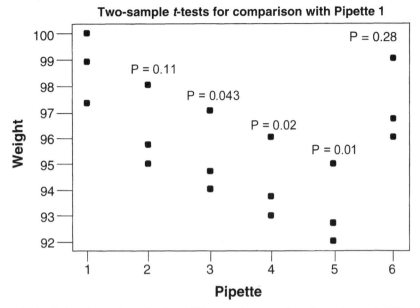

Figure 4.17 Point plots of manipulated *Wt mg* values, to show how the probability (*P*) in the *t*-test corresponds to the extent of vertical separation between Pipette 1 and Pipettes 2–6

The next three entries in each column were produced by a similar copy and paste from Pipette 2, followed by manipulations. In Col. 5, 1 mg was subtracted from each Pipette 2 value so as to make all the deliveries from that pipette (now called Pip 3) smaller by that amount. In Col. 6 the difference was 2 mg (to give Pip 4) and in Col. 7, 3 mg (to give Pip 5). In Col. 8 the two pipettes were made closer together by *adding* 1 mg to each original Pip 2 value (to give Pip 6). Figure 4.17 presents point plots of the manipulated data sets, using Cols 3 and 4 for this purpose. Pipettes 1 and 2 are the unaltered data, while 3, 4, 5 and 6 are the manipulated groups.

The *P*-values on Figure 4.17 are those generated by MINITAB as above from the pairs of pipettes in Cols 2, 5, 6, 7 and 8. They show that the t-test on Pip 3 and Pip 1, where there is no overlapping of points, produces a significant difference (*P* = 0.043). When a further 1 mg was removed to give Pip 4, the *P*-value fell to 0.02, and with a 3 mg difference in Pip 5, the difference from Pip 1 was highly significant (*P* = 0.01). The final set of Pip 6, which overlapped strongly with Pip 1, had a higher *P* = 0.28 than that given by the original two-pipette data set (Pip 1 and Pip 2) without manipulation.

When ANOVA was applied to the manipulated data sets, the *P*-values were virtually coincident with the *t*-test results: viz Pip 1 and Pip 3 gave *P* = 0.043; Pip 1 and Pip 4 gave *P* = 0.02; Pip 1 and Pip 5 gave *P* = 0.01; and Pip 1 and Pip 6 gave *P* = 0.279.

4.6 SUMMARY

This chapter has continued the study of the student pipetting results as an example of normally-distributed measurement data. The emphasis has been on initial visual inspection and interpretation of the data by *point plots and boxplots*. This was followed by showing how the initial visual impressions about differences in group means may be confirmed by analysis of variance and *t*-test. No attempt was made to work with equations or statistical tables, since MINITAB does not display these as part of its output. Thus the material presented has been introductory and practical, rather than theoretical and mathematical. A further limitation is the small part of MINITAB's resources for analysis of variance that was used.

5

Awkward-Measurement Data

The statistician must be treated less like a conjurer whose business is to exceed expectation, than as a chemist who undertakes to assay how much of value the material submitted to him contains.

R. A. Fisher (1890–1962)

5.1 AWKWARDNESS IN MEASUREMENT DATA

The topics so far discussed in this book have been restricted in three ways:

- The data have consisted of *measurements*.
- They have been a reasonably close approximation to a *normal distribution*.
- When several groups were involved, they all had approximately the same amount of internal scatter (st. dev.).

Thus the pipetting results of Chapters 1–4 were treated as random samples from one or more 'underlying' normal distributions with similar st. devs. This allowed the use of an extensive set of statistical tools for display, summary and analysis, the pipetting data being taken as a convenient example of many types of measurement data. These chapters, in statisticians' language, are concerned with the *normal* model with *uniform variances*. More on this 'model' follows in the later chapters on Correlation and Regression (Chapter 8) and Dose–Response Lines and Assays (Chapter 9).

Unfortunately in the 'real world' (as they say), not all types of measurement data can be treated in the same way as the pipette deliveries. In particular, the data may contain one or more of the awkward features that are the subject of the present chapter and which may frustrate use of the previously described statistical tools. For convenience, the different types of awkwardness are presented individually, but in real life they may occur all together in the same data set. This area is one where the advice of a professional statistician should be sought early, in order to avoid pitfalls and to make the best use of the data. Therefore one of the purposes of this chapter is to show some samples of 'awkwardness' in measurement data, so that a higher level discussion with a statistician can take place.

5.2 STANDARD DEVIATIONS GROSSLY DIFFERENT

It is common in physiological experiments to have significantly different standard deviations, for example in *Treatment* and *Control* groups. A sample of such data is provided in Table 5.1, which shows the concentration of white blood cells (wbc) in the blood of 24 control rats and of 16 animals infected in their lungs with pertussis bacteria. These data could also be considered as *counts* and therefore better placed in the next chapter. In a purist sense, at the quantal level, all measurement data could be considered as counts. For example, the *Wt mg* data of the pipetting experiment could be considered as *counts of tenths of a milligram* since the electronic balance is digital on this scale. All of this shows that excessive semantic rigidity may be unhelpful.

The lower section of Table 5.1 displays a selection of the *summary statistics* that show the infected animals to have about three times the concentration of wbc as the controls, but to have a st. dev. that is 12 times that of the controls. These features are well displayed by the point plots and boxplots of the data in Figure 5.1, which show that the st. devs of the two groups are grossly different (as are also the means, but that does not matter, from a validity standpoint).

Table 5.1 White blood cell (WBC) concentrations (mm^{-3}) sorted in ascending order in each group, and summary statistics, on 24 control rats and 16 rats that had been infected with pertussis bacteria

Control	Infected	Control	Infected
10252	11020	12088	48272
10467	14181	12586	49062
10601	16753	12693	58885
10638	16907	12847	69916
10901	28898	12850	
11071	29526	13096	
11092	31801	13161	
11371	35445	13527	
11667	37423	13560	
11686	38812	13720	
11688	39243	14601	
11860	45869	15159	

Selected Summary Statistics

Group	N	Mean	Median	St. dev.	SE mean	95.0% CI
1 (Control)	24	12216	11974	1335	273	(11652, 12780)
2 (Infected)	16	35751	36434	16447	4112	(26987, 44515)

5.2.1 MINITAB worksheet

To investigate the statistical properties of the *WBC Conc* data, the concentrations were entered on a MINITAB worksheet, as in Table 5.2. Column C1 contains the *Group* code of 1 = *Control* rats and 2 = *Infected* animals, while C2 details the wbc concentrations (*WBC Conc*). In addition to having the data in the format of Cols C1 and C2, the wbc concentrations in the two groups are also listed in individual columns on the worksheet. C3 (*Conc Cont*) contains the wbc concentrations in the *Control* rats, and C4 (*Conc Inf*) the values for the *Infected* animals. These were produced by MINITAB's *Unstack Columns* feature (below).

Previously described methods were then used to obtain the summary statistics (§1.8.3, §2.3.1), 95% confidence intervals (§2.5.3) point plots (§1.7, §4.1.1) and boxplots (§4.1.2, §4.1.3). In each of these, *WBC Conc* was taken as the *variable*, or *Y-axis* entry, and *Group* as the *factor*, or *X-axis* entry.

5.2.2 Unstacking columns

For the Mann-Whitney test (below), it is necessary to have the data from the *Control* and *Infected* animals in separate columns, whose headings should be entered in C3 and C4 on the worksheet as shown. After this, go to **Manip > Stack/Unstack > Unstack One Column**, and when the dialog box (Figure 5.2) opens:

1) At $\boxed{\text{Unstack the data in:}}$, enter '*WBC Conc*';
2) At $\boxed{\text{Store the unstacked data in:}}$, enter '*Conc Cont*' '*Conc Inf*';
3) At $\boxed{\text{Using subscripts in:}}$, enter *Group*;
4) Click on $\boxed{OK}$.

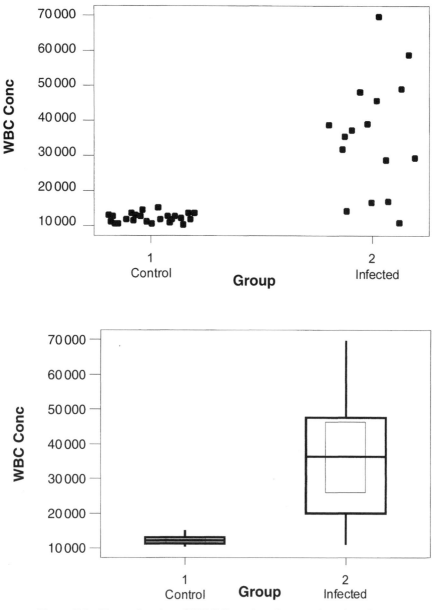

Figure 5.1 Top, point plot of *WBC Conc* data; bottom, boxplot of same

Confirm that this has unstacked C2 and delivered its contents correctly to C3 and C4.

5.2.3 Inequality of standard deviations

The specific test for exploring whether the st. devs of the two groups of rats are different is that for *homogeneity of variances,* the *variance* being the square of the

Table 5.2 MINITAB worksheet for analysis of the rat
wbc data

C1 Group	C2 WBC Conc	C3 Conc Cont	C4 Conc Inf
1	10 252	10 252	11 020
1	10 467	10 467	14 181
1	10 601	10 601	16 753
1	10 638	10 638	16 907
1	10 901	10 901	28 898
1	11 071	11 071	29 526
1	11 092	11 092	31 801
1	11 371	11 371	35 445
1	11 667	11 667	37 423
1	11 686	11 686	38 812
1	11 688	11 688	39 243
1	11 860	11 860	45 869
1	12 088	12 088	48 272
1	12 586	12 586	49 062
1	12 693	12 693	58 885
1	12 847	12 847	69 916
1	12 850	12 850	
1	13 096	13 096	
1	13 161	13 161	
1	13 527	13 527	
1	13 560	13 560	
1	13 720	13 720	
1	14 601	14 601	
1	15 159	15 159	
2	11 020		
2	14 181		
2	16 753		
2	16 907		
2	28 898		
2	29 526		
2	31 801		
2	35 445		
2	37 423		
2	38 812		
2	39 243		
2	45 869		
2	48 272		
2	49 062		
2	58 885		
2	69 916		

st. dev. To do this on MINITAB, go to **Stat > ANOVA > Homogeneity of Variance**
and, when the dialog box (Figure 5.3, top) appears:

1) At $\boxed{Response:}$, enter '*WBC Conc*';
2) At $\boxed{Factors:}$, enter *Group*;
3) Click on $\boxed{OK}$.

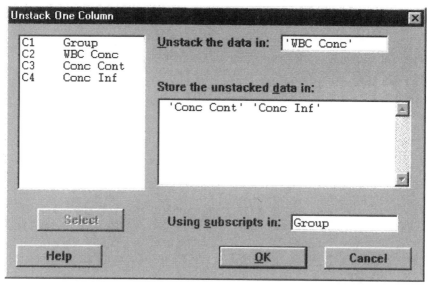

Figure 5.2   MINITAB dialog box for unstacking a column of data

The output, in the lower part of Figure 5.3, presents first the 95% confidence intervals of the st. devs (referred to as 'Sigmas') for the two groups (of wbc concentrations), called 'factor levels'. The wide separation of the horizontal plots of the CIs indicates a highly significant difference in the st. devs. This is confirmed in the numerical output on the right where the *F*-test statistic is given as 151.723. This statistic is the ratio of the two st. devs squared, i.e. with the values from Table 5.1:

$$F = (\text{st. dev.}_1/\text{st. dev.}_2)^2 \qquad\qquad \text{Eq. 5.1}$$
$$= (16\,447/1335)^2$$
$$= (12.3198)^2$$
$$= 151.778, \text{ as given in Figure 5.3.}$$

This is associated with a null hypothesis probability of $P = 0.000$, which indicates a highly significant difference between the st. devs of the *Control* and *Infected* rats. A similar result emerges from the difference between the boxplots at the bottom of the figure and its associated statistics. For information on Levene's test, see MINITAB *User's Guide 2*, p. 5–12.

5.2.4 Mann-Whitney Test

An important consequence of the significant difference in st. devs is the *invalidation* of the *t*-test to explore possible differences in group means. This is because one of the requirements for a valid *t*-test is that the standard deviations of the two groups being analysed should not be significantly different. An alternative test in such situations is the Mann-Whitney Test, which does not require the data to have similar st. devs, nor even to be from underlying normal distributions. Mann-Whitney is an example of a non-parametric test, since no assumptions are made about the nature of the underlying distribution, that is, the existence of parameters

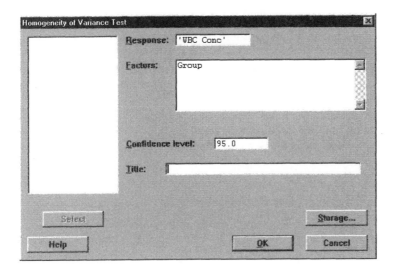

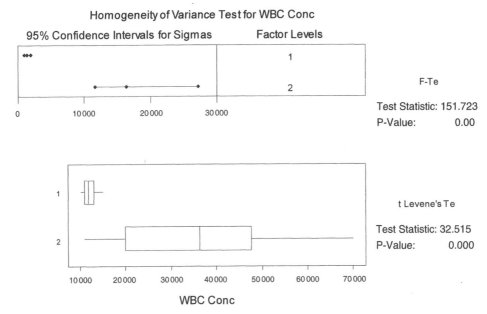

Figure 5.3 Top, MINITAB dialog box for *Homogeneity of variance test*; bottom, output of the test on the *WBC Conc* data

like mean and st. dev. are not even considered, let alone assumed. Thus instead of the mean being taken as the measure of central tendency, the median is used.

The Mann-Whitney test itself takes the *ranks* of the measurements, the two groups being considered jointly for purposes of ranking. Thus the lowest measurement in the whole set of *WBC Conc* values has a rank of 1, the next a rank of 2, and so on. Allowance has to be made for identical measurements having joint ranks. The mechanics of the test involves calculating a test statistic W by a complex procedure that is outlined in the MINITAB *User's Guide*, and determining the significance of the difference between the medians of the two groups. The Greek letter η (eta) is used for the medians, so there is η_1 for the median of Group 1 and η_2 for Group 2. The crucial feature is whether the 95% confidence intervals of $(\eta_1 - \eta_2)$ *enclose* zero. If they *do*, the medians of the two groups are not significantly different; conversely if the confidence limits do not enclose zero, then the differences *are* significant.

To do the Mann-Whitney test on MINITAB requires that the two groups of data are in separate columns, which is why they were unstacked above. With this in place, go to **Stat > Nonparametrics > Mann-Whitney** and, when the dialog box (Figure 5.4, top) appears:

---

1) At `First Sample:`, enter '*Conc Cont*' for the Control wbc results;
2) At `Second Sample:`, enter '*Conc Inf*' for the Infected wbc results;
3) Click on `OK`.

The output appears in the Session window as in Table 5.3, where line numbers have been added to assist discussion. Lines 1 and 2 identify the two groups of observations, the numbers (*N*) in each group, and their medians, 11 974 for the *Control* group and 36 434 for the *Infected* animals.

- Line 3 states that the point estimates for the difference between the medians (eta values) is –24 147 (which is not the simple arithmetic difference). The negative sign should be ignored since it was arbitrary how the groups were identified as 1 and 2. Thus if the *infected* group had been labelled as No. 1, the difference in medians would have been positive. The null hypothesis states that the difference in medians is zero, i. e. that ETA 1 = ETA 2.
- Line 4 gives the confidence intervals of the difference in the ETA values as −28 991 and −17 038. Neither of these overlaps with zero, which indicates that the difference in medians is significant. Note that the CIs are described as 95.2% CIs, instead of the usual 95%. This is due to exact 95% limits not being available from ranked data.
- Line 5 gives the Mann-Whitney statistic *W*, whose value is only meaningful in the context of the equation that produced it. Technically, *W* is the sum of the ranks of the observations in the first group (see MINITAB *User's Guide 2*, p. 5–12).

Table 5.3 MINITAB output for the Mann-Whitney Test on the WBC concentrations in the blood of *Control* and *Infected* rats. Line numbers (1), etc., have been added to assist discussion

---

Mann-Whitney Confidence Interval and Test

1) WBC Con   N = 24   Median =   11 974
2) WBC Inf   N = 16   Median =   36 434
3) Point estimate for ETA1 – ETA2 is –24 147
4) 95.2 Percent CI for ETA1 – ETA2 is (−28 991−17 038)
5) W = 321.0
6) Test of ETA1 = ETA2 vs ETA1 not = ETA2 is significant at $P = 0.0000$

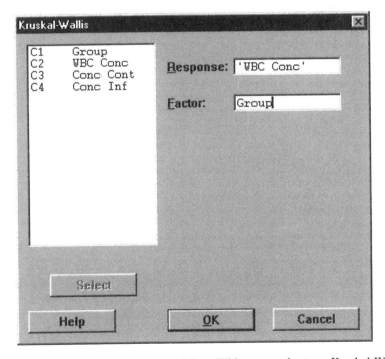

Figure 5.4 MINITAB dialog boxes. Top, Mann-Whitney test; bottom, Kruskal-Wallis test, on *WBC Conc* data

- Finally, the *P*-value for the null hypothesis of the two median*s* being the same, is 0.0000, which is very highly significant. Therefore the null hypothesis can confidently be rejected and the medians of the two groups of wbc counts declared as being highly significantly different.

Thus the Mann-Whitney provides a very useful 'fallback position' for comparing the median*s* of two groups of data when the *t*-test is inapplicable because of significant differences in the st. devs.

5.2.5 Non-parametric 'analysis of variance'

The Mann-Whitney test is for comparing the medians of two groups. If there are more than two groups, there is a non-parametric equivalent of analysis of variance that can be done. This, the Kruskal-Wallis test, determines the significance of differences of the medians of grouped data, from two groups upwards. Unlike Mann-Whitney, it requires the observations to be presented in a single column, with the group designation as code numbers alongside. This is already available on the MINITAB worksheet as C1 and C2 in Table 5.2. To apply this test, go to **Stat > Nonparametrics > Kruskal-Wallis** and, when the dialog box (Figure 5.4, bottom) appears:

---

1) At Response: , enter 'WBC Conc';
2) At Factor: , enter Group;
3) Click on OK .

Table 5.4 MINITAB output for the Kruskal-Wallis Test on the WBC concentrations in the blood of *Control* and *Infected* rats. Sections A and B have been added for reference in discussion

Kruskal-Wallis Test on WBC Conc				
Section A				
Group	N	Median	Ave Rank	Z
1	24	11 974	13.4	−4.72
2	16	36 434	31.2	4.72
Overall	40		20.5	
Section B				
H = 22.29	DF = 1	P = 0.000		

The output in the Session window is as in Table 5.4. Section A summarizes the main features of the data, the number (*N*) observations in each group, the average rank of each group, and of the whole data set, and a quantity *Z* that indicates how the mean rank of the group differs from the mean rank of all the observations. The exact technical definition of *Z* is given in the MINITAB *User's Guide*.

In Section B is given the Kruskal-Wallis test statistic, *H*, which can be approximated to by a chi-square distribution (see FN 6.2) with $(k - 1)$ degrees of freedom, where *k* is the number of groups. A significant result at the $P = 5\%$ level requires a chi-square of ≥ 3.84. Therefore the value of $H = 22.29$, with one degree of freedom

(DF = 1), is associated with a very low probability ($P = 0.000$). Thus the difference in medians is, as in the Mann-Whitney test, very highly significant.

As with Mann-Whitney, Kruskal-Wallis may come to the rescue, as an alternative to ANOVA, with data sets that fail the test for homogeneity of variances.

5.3 SKEWED DISTRIBUTIONS

5.3.1 Antibody titres

Another departure from the the *normal* model with *uniform variances* is where the distribution is *skewed*. This is shown by the antibody titre data of Table 5.5 and displayed in the top part of Figure 5.5 as a histogram with superimposed distribution curve. The titres range over the extremely wide range of 1100 to 550 000 that is a characteristic of antibodies and which reflects the high variability in responsiveness of the individuals (in this case rats) who receive immunization with a vaccine. This output was obtained as in §4.2.2 by going to **Stat > Basic Statistics > Display Descriptive Statistics** and, when the dialog box appeared, entering 'Anti-FHA' in *Variables*. The internal sub-box for *Graphs* was opened and ☑ *Histogram of data with normal curve* chosen.

Table 5.5 Concentration of anti-FHA (antibodies to filamentous haemagglutinin) in the serum of 31 vaccinated subjects arranged in ascending order. The results are expressed as reciprocal ELISA* titres

1100	6900	13 500	29 000	80 000
1600	7000	16 000	34 000	85 000
2200	7500	19 000	50 000	100 000
4100	7500	20 000	52 000	100 000
6500	7900	23 000	60 000	370 000
6500	8500	25 000	65 000	550 000
6500				

*Enzyme-linked immunosorbent assay.

5.3.2 Logarithmic transformation

A well-established procedure for dealing with such results, statistically, is to perform a logarithmic transformation, which has the effect of converting the skewed distribution into a symmetrical one (Figure 5.5, bottom). Data that are made symmetrical in this way are described as *lognormal*. The worksheet (not shown) started with the plain titres as a list of 31 values in C1 labelled Anti-FHA, and an empty column C2, labelled Log AntiFHA. To make the transformation, open **Calc > Calculator** and, when the window (Figure 5.6) appears:

1) At Store result in variable:, enter '*Log AntiFHA*';
2) At *Expression*, enter LOGT('Anti-FHA'), making use of the *Functions* menu to choose Log 10, which becomes LOGT in *Expression*;
3) Click on OK.

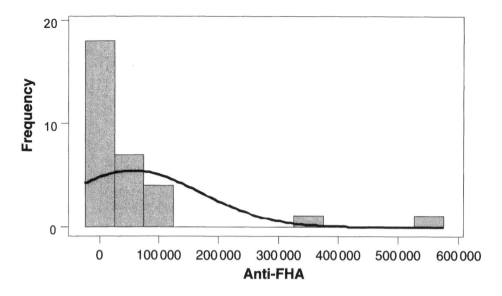

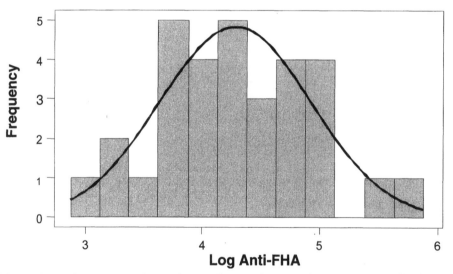

Figure 5.5 Histograms and superimposed normal curves for *anti-FHA* antibody titres. Top, original data; bottom, after logarithmic transformation

The column of Log_{10} values was then treated in the same way as the initial data to give the symmetrical distribution in the lower diagram of Figure 5.5.

5.3.3 Descriptive statistics

For summary purposes, the mean and either the SE mean or the 95% confidence intervals will be required. As part of output of the requested descriptive statistics,

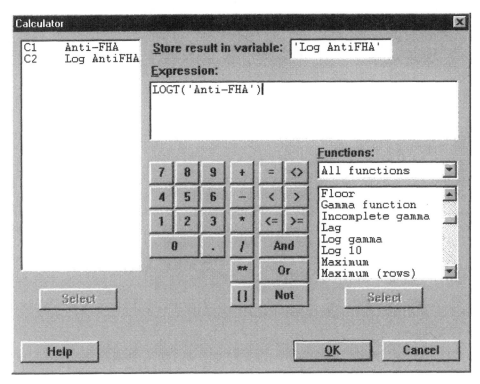

Figure 5.6 MINITAB dialog box for log transformation of *anti-FHA* titres

MINITAB delivered to the Session window the mean, st. dev. and SE mean as Log_{10} values. However, the 95% CIs have to be obtained separately by going to **Stat > Basic Statistics > 1-Sample t** and when the dialog box (Figure 5.7) appears:

1) At $\boxed{Variables}$, enter '*Log AntiFHA*';
2) Leave the default options unchanged and click on $\boxed{OK}$.

This delivers to the Session window the statistics wanted as:

Variable	N	Mean	St. dev.	SE Mean	95.0% CI
Log Anti	31	4.281	0.641	0.115	(4.046, 4.516)

With a pocket calculator, taking the antilogs gives:

- Mean = 19 098.533, which may be rounded to 19 000; and
- 95% CI = 11 000 and 33 000.

The antilog of the mean logarithm, i.e. the 19 000, is also known as the geometric mean. Taking antilogs of the st. dev. and SE mean should be avoided, as they are only valid parameters for the symmetrical distribution with the logarithmic X-axis. If one wanted to have the range of mean $\pm$ 1 SE mean, the calculation should be done in logs and then the antilogs applied. So that this expression in logs would be

4.281 ± 0.115 = 4.166 and 4.396. Now taking antilogs gives the mean ± 1 SE mean as 14 655 and 24 888, which may be rounded to 15 000 and 25 000, considerably narrower than the confidence intervals.

5.3.4 Test for normality

To confirm that the original data were lognormally distributed and that the logarithmic transformation 'normalized' them, the methods used in §3.4 were applied. When the normality test is done on the original anti-FHA titres, the probability plot (Figure 5.8, top) has the experimental points on a curve and the fitted straight line largely independent of them. At the bottom right of the figure, the Ryan-Joiner test has a *P*-value <0.01, indicating a highly significant departure from normality.

After the logarithmic transformation, however, the experimental points are randomly scattered around the independently-fitted straight line, and the Ryan-Joiner *P*-value >0.1 indicates that departures from normality are not significant. Thus the logarithmic transformation was successful in normalizing the anti-FHA titres. Such normalized data would now be suitable for the application of normal distribution statistics as were the pipetting data.

5.4 DATA SETS WITH INDEFINITE VALUES

5.4.1 Undetectable titres

In the ELISA procedure for assaying antibodies, there may be no detectable antibody at the highest concentration of serum tested (Table 5.6). Such a result can not accurately be reported as a titre of *zero* because the actual result is in the form

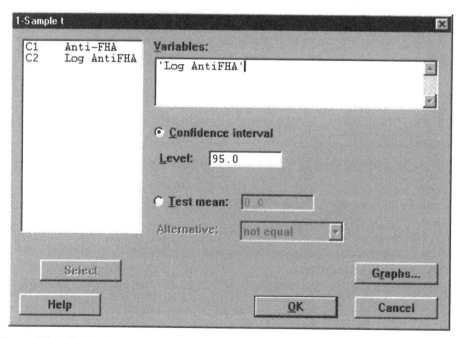

Figure 5.7 MINITAB dialog box for obtaining 95% confidence intervals of the mean *log anti-FHA* titre

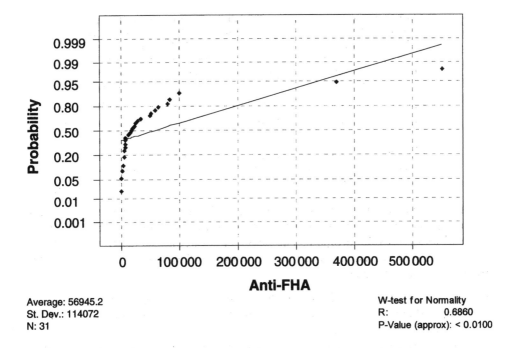

Average: 56945.2
St. Dev.: 114072
N: 31

W-test for Normality
R: 0.6860
P-Value (approx): < 0.0100

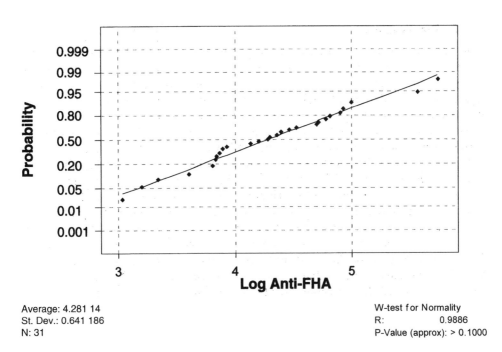

Average: 4.281 14
St. Dev.: 0.641 186
N: 31

W-test for Normality
R: 0.9886
P-Value (approx): > 0.1000

Figure 5.8 Probability plots for normality test of *anti-FHA* titres. Top, original data; bottom, after logarithmic transformation

Table 5.6 Concentration of anti-PT (antibodies to pertussis toxin) in the serum of 30 vaccinated subjects, arranged in ascending order. The results are expressed as reciprocal ELISA* titres

<100	<100	1000	1300	11 000
<100	<100	1000	1660	19 000
<100	<100	1000	1800	24 000
<100	<100	1000	2900	40 000
<100	1000	1000	3400	44 000
<100	1000	1000	5000	57 000

*Enzyme-linked immunosorbent assay.

of (say) a titre of <100. This presents a problem with MINITAB since such data can not be entered on the worksheet as valid numerical information. There is probably no *good* solution to the problem. The question is whether there are ways forward that are *acceptable*. Simply to enter the number '0' wherever there is a '<100' creates several difficulties:

- It may exaggerate the extent to which the true titre is less than 100;
- It would therefore give a false low value for the geometric mean titre of the whole set;
- It would prevent a logarithmic transformation, since the log of zero is indeterminate.

5.4.2 Practicalities

One possibility is to substitute an arbitrary number of, say 50 as a *notional* titre wherever there is a '<100'. This would be declared as a procedural step in the analysis, and would be prevented from causing false outcomes by excluding such figures from calculations of means or geometric means. Instead the median would be used as the measure of central tendency, as this would not be affected by arbitrary values replacing '<100'. This has therefore been done in C1 of a worksheet (not shown) similar to that used for anti-FHA. C2 of the worksheet had the logs of the titres, as before. The histograms of anti-PT before and after the log transformation are shown in Figure 5.9 and are not unlike those for anti-FHA in Figure 5.5.

Likewise, the probability plots (Figure 5.10) suggest that anti-PT was lognormally distributed, since the Ryan-Joiner statistic became non-significant (P >0.1) after the log transformation. It would however be misleading to go on and calculate the geometric mean and confidence intervals, because of the fictitious values used in place of the <100 titres. In the circumstances, the prudent action would be to calculate the median and its confidence intervals. This can be done on the untransformed data by going to **Stat > Nonparametrics > 1-Sample Sign** and when the dialog box (Figure 5.11) appears:

1) At ⎹ *Variables* ⎸, enter '*Anti-PT*';
2) Leave the default options unchanged and click on ⎹ *OK* ⎸.

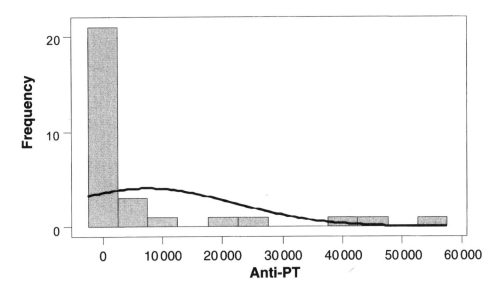

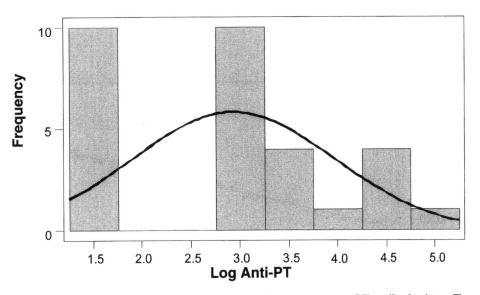

Figure 5.9 Histograms and superimposed normal curves for *anti-PT* antibody titres. Top, original data; bottom, after logarithmic transformation

The output in the Session window is:

	N	Median	Confidence achieved	Confidence intervals	Position
Anti-PT	30	1000	0.9013	(1000, 1660)	11
			0.9500	(267, 1768)	NLI
			0.9572	(50, 1800)	10

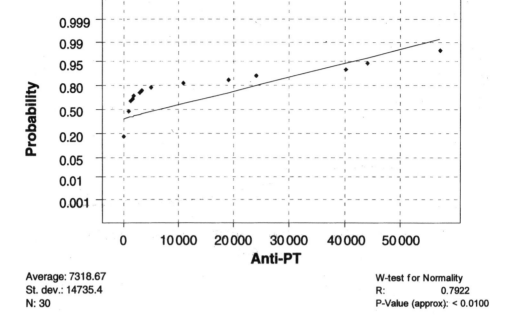

Average: 7318.67
St. dev.: 14735.4
N: 30

W-test for Normality
R: 0.7922
P-Value (approx): < 0.0100

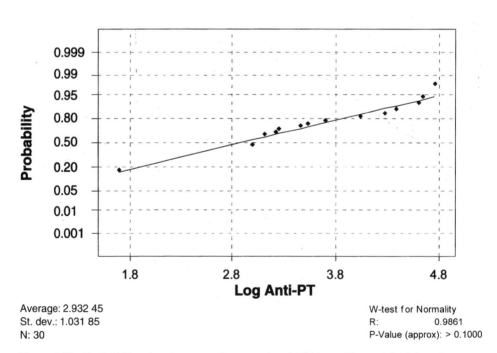

Average: 2.932 45
St. dev.: 1.031 85
N: 30

W-test for Normality
R: 0.9861
P-Value (approx): > 0.1000

Figure 5.10   Probability plots for normality test of *anti-PT* titres. Top, original data; bottom, after logarithmic transformation

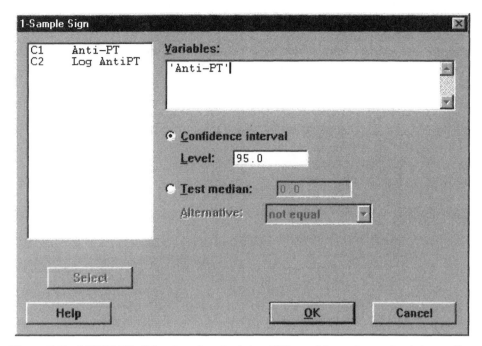

Figure 5.11 MINITAB dialog box for obtaining 95% confidence intervals of the median *anti-PT* titre

This states that with the $N = 30$ observations of anti-PT, the median titre is 1000, with 95% confidence intervals of 267 and 1768 obtained by non-linear interpolation (NLI). The other CIs are for 90.13% and 95.72% which are not of interest.

The $N = 30$ anti-PT titres could thus be summarized as having a:

- *median* titre of 1000, with 95% CIs of 300 and 1800
- *Range* from <100 to 57 000
- 10/30 of the immunized subjects showing no response at the limit of detection (titre = 100).

The existence of 'non-responders' is a well-established phenomenon in immunology, but these subjects would require further study before confident assignment to such a category. If several sets of anti-PT titres were to be analysed, the task would best be worked at by non-parametric methods such as Mann-Whitney and Kruskal-Wallis and using the 1-sample sign test for confidence intervals.

5.5 SUMMARY

This chapter has taken three sets of measurement data that present problems in their statistical analysis. The first set had data where the st. devs of different groups were very different from each other. This did not prevent the use of the mean, st. dev. and SE mean as valid descriptive statistics, but did negate the use of the *t*-test and ANOVA for exploring differences in group means. Instead the analysis of group median*s* was done by non-parametric methods, namely the Mann-Whitney and Kruskal-Wallis tests.

In the second set, the underlying distribution was skewed, but this could be resolved by a logarithmic transformation, which then allowed application of normal distribution statistics. The third data set had '<', values which could not be used as numerical input to MINITAB. This problem was dealt with by assigning notional values instead of the '<100' titres, and avoiding false-output statistics by dealing only with medians and their confidence intervals. Although these examples of 'awkwardness in measurement data' were considered separately, there is the possibility of them occurring together. The advisability of seeking expert statistical advice was emphasized.

6

How to Deal with Count Data

Truth is rarely pure and never simple.

Oscar Wilde (1854–1900)

6.1   FROM BACTERIAL COLONIES TO UMBRELLAS LEFT ON BUSES

After five chapters devoted to *measurement* (also known as *continuous*) data, the theme now changes to *discontinuous*, or *count*, data (also known as *discrete* data). The next chapter is also devoted to discrete data, but of the kinds that consist of *proportions*, such as 3/11 for the proportion of grant applications that were successful. Such proportion data are generated in the counting of microorganisms by the most probable number (MPN) method and are also in the next chapter.

Just as with measurement data, where there was an underlying idea (not necessarily correct!) that the normal distribution would be the basis for statistical processing, so with count data the first focus will be the Poisson distribution (which, again, is only correct in some instances). Poisson is introduced with the small diversion of asking what is the common factor in the following?

- Bacterial colony counts per culture plate;
- Number of radioactive disintegrations per second;
- Number of umbrellas left per day on the buses in a large city;
- Number of flying bomb hits per square kilometre of London during the Second World War;
- Number of men kicked to death by horses each year in the Prussian army in the days before motorized transport;
- Death notices per day for men over 85 in the obituary columns of the London *Times*;

Answer: the common feature in all of these, from a statistical standpoint, is that:

1) The data consist of *whole-number* observations, i.e. *counts*. There can not be 4.61 colonies on a culture plate, or 4.61 umbrellas left on city buses yesterday.
2) Each object or event that is recorded is a very small fraction of the total number 'at risk' (to use the jargon phrase). The bacterial colonies on the plate originated from a very small proportion of the total number of live organisms in the bulk suspension ('at risk' of being plated on to that culture plate). Likewise the lost umbrellas were a very small proportion of the total number being carried that day on the buses (and 'at risk' of being parted from their owners).
3) The objects or events are *independent* of each other and occur randomly, in that the leaving of one umbrella on a bus does not influence the leaving of other umbrellas on other buses.
4) The number of objects or events counted in each unit of space, volume or time is usually small. This is true of bacterial colony counts if the dilution is sufficiently large and of radioactivity counts if the time interval is sufficiently short. Here 'small', means less than about 10.

6.1.1   Introduction to the Poisson

The Poisson distribution, or Law of Small Probabilities, is a concept that applies to the *counts* of objects or events that are randomly distributed in area, volume or time, and where a large number of them is 'at risk', but only a small proportion is 'captured' within each observational unit of area, volume or time.

There are four main points about the Poisson distribution:

- It describes at least approximately many types of count data generated by the experimental biologist, such as bacterial colony counts, virus plaque counts, total cell counts and radioactivity counts.
- It provides a theoretical basis for analysing a variety of one-hit phenomena such as phage–bacteria and radiation–genome interactions;
- If the count per unit of volume or time is sufficiently large, the Poisson distribution approximates to a normal distribution;
- Living organisms in their natural environments are *unlikely* to be Poisson distributed, because they tend to live in clusters of some kind, rather than being randomly distributed

This chapter explores some of the common aspects of count data and of the Poisson distribution that are likely to be needed by the experimental biologist. Also to be introduced is the chi-square test that is used to analyse differences between counts, in the same way as the *t*-test and ANOVA are employed to assess differences between means.

6.1.2 Limitations of MINITAB

MINITAB does so many useful and sophisticated calculations that one may be surprised to find some simple things that it does not do. Several such surprises occur in this chapter, an example being the lack of an inbuilt MINITAB function to obtain the confidence intervals of a count. Likewise there is no inbuilt function to explore directly the significance of differences between two or more counts. Such shortcomings can, of course, be remedied by writing appropriate MINITAB 'macros' but this has not been done here. Instead the approach is to use the provided functions on MINITAB where possible, and to adopt a formula-and-calculator procedure when required. It is noteworthy that the words *count* or *counts* are not in the index of the MINITAB Inc. (1997a, b) *User's Guides*, or as subheadings under *confidence intervals*. Another feature of this chapter is to display fewer of the MINITAB dialog boxes than heretofore, on the assumption that the reader will be able to follow the box dialog from the text.

6.2 TOTAL CELL COUNTS AS A WORKED EXAMPLE

6.2.1 Cell suspensions in a counting chamber

Before the widespread availability of the Coulter Counter, it was commonplace (and still is) to do counts of the total number of cells (bacterial, algal, blood, etc.) per unit volume of a suspension, by microscopy. The counting is from special microscope slides with grid lines engraved on a depressed area of the surface. One such device is the Helber slide counting chamber, which is a thicker than normal microscope slide with a central, optically-flat platform that is an accurate 0.002 cm below, and parallel to, the rest of the surface. Engraved on this platform are criss-crossing grid lines that form a pattern of small squares of side 0.005 cm. A coverglass placed on the walls bounding the sunken platform creates a wafer-thin chamber into which a cell suspension can be introduced by capillary action. The volume of fluid above each

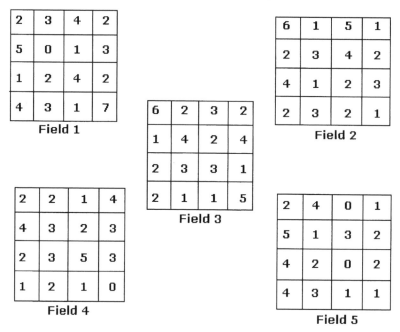

Figure 6.1   Record of the distribution of *E. coli* cells over the 80-square pattern of grid lines in a Helber slide counting chamber

small square in the chamber is then $0.002 \times 0.005 \times 0.005$ cm^3, or 0.5×10^{-7} cm^3. If the subsequent average count of cells is 1.0 per small square, the count per cm^3 (equivalent to per ml) in the suspension is $(1/0.5) \times 10^{-7}$, or 2.0×10^7 ml^{-1}.

The small squares are set out in square arrays of 16 which, in turn, are in clusters of five, so as to give a set of 80 squares in a diagonal cross pattern as shown in Figure 6.1.

The usual procedure is to count the number of cells floating above the 80 squares under appropriate magnification and using a hand-held tally counter, and then dividing the total count by 80 to obtain the average count per square. The counting is carried out in such a way that cells lying on a grid line are assigned un-ambiguously to the square underneath or to the left of the line, thereby avoiding either counting the same cell twice or omitting it. Multiplication of the average count per square by 2.0×10^7 gives the count per ml in the bulk suspension. It is also customary to empty and refill the chamber with further portions of the same suspension so as to obtain duplicate or triplicate estimates of the average count. As a non-statistical aside, it may be noted that the Helber chamber is only suited for suspension with counts in the range 10^7–10^8 ml^{-1}. However, other counting chambers with larger dimensions, and correspondingly larger above-square volumes allow the counting of less dense suspensions.

Although the Helber slide is to be used here as a specific example, its generality should also be appreciated. For example, the grid squares can be regarded as equivalent to the square kilometres of London on which flying bombs fell; or as buses on which umbrellas were left; or as units of time representing the days on which obituary notices appeared in the London *Times*; or as Petri dishes with bacterial colonies; or as hundredths of a second in which radioactive disintegrations occurred. All are examples of Poisson distributed counts.

Likewise the five *fields* of the Helber data have their counterpart in, for example, different Petri dishes in which bacterial colonies (from the same plating) were growing; or different months of the year in which umbrellas were left on city buses; or different areas of London on which flying bombs fell. These *fields* thus constitute groups or subsets within the whole assemblage of observations, which can be tested for *homogeneity*.

Unless there is a special reason, as here, to record the counts in the individual squares of the Helber chamber, the normal procedure would be simply to obtain the total count over the 80 squares and calculate the average count per square. However, to explore the way the cells are spread, in terms of a Poisson distribution, it is necessary to record the number of cells in each square as in Figure 6.1. This was from a suspension of *E. coli* cells, prepared by making a 1 : 20 dilution of an overnight broth culture into formalin saline.

6.2.2 Preliminary processing without MINITAB

For instructional purposes preparatory to MINITAB processing, it is useful to perform a pencil-and-paper analysis of the data in Figure 6.1 to reveal the underlying frequency distribution. The critical variable is the *occupancy number* (here given the symbol x) of the grid squares. These occupancy numbers consist of the whole-number values from zero up to the maximum count in an individual square, which in this instance is seven. They make up the first column of Table 6.1. Adjacent to them are the pencil tick marks that were recorded as each square of Figure 6.1 was inspected and its count noted, when the fields were scanned systematically from left to right and top to bottom. At this stage turning Table 6.1 through 90° reveals a slightly skewed distribution with a peak in the $x = 2$ category.

In the third column of Table 6.1 the ticks for each x-value are added, and in the final column they are expressed as a percentage. Arithmetical check on the

Table 6.1 Frequency distribution of the bacterial cells over the grid squares in the Helber slide counting chamber (summarized data from Figure 6.1)

Occupancy number of the square (x)	Tally of squares with indicated number of bacterial cells	Total number of squares	Percentage of squares
0	////	4	5.0
1	///// ///// ///// ///	18	22.5
2	///// ///// ///// ///// ///	23	28.75
3	///// ///// /////	15	18.75
4	///// ///// //	12	15.0
5	/////	5	6.25
6	//	2	2.5
7	/	1	1.25
	Total	80	100.00

summations confirms that the numbers in columns 3 and 4 add up to 80 and 100 respectively.

6.2.3 MINITAB worksheet for Helber counts

For the processing of these data with MINITAB, open a new MINITAB project and, on the worksheet, label the first two columns *Field* and *Count*. The *Field* column needs a set of 80 numbers made up of 16 repetitions of the numbers 1 to 5 and may be done with the *Make Patterned Data* command:

Go to **Calc > Make Patterned Data > Simple Set of Numbers**, and when the dialog box opens, fill the spaces as:

---

1) *Store patterned data in:* ⌑Field⌑;
2) *From first value:* ⌑1⌑
3) *To last value:* ⌑5⌑
4) *In steps of:* ⌑1⌑
5) *List each value:* ⌑16⌑ *times;*
6) *List the whole sequence:* ⌑1⌑ *times;*
7) ⌑*OK*⌑.

This will give the sets of 16 repetitions for each *Field*, as in Table 6.2. The C2 column in this table is then copied in from the counts in Figure 6.1 and afterwards checked for accuracy of recording. Having completed Cols 1 and 2, the project should be saved as *Chap 6 Helber.mpj*.

Next, it is useful to *sort* the counts in numerical order, as shown in Col. 3 of Table 6.2. Therefore label Col. 3 as *Sorted count*. To allow a caption with this number of characters, expand the column width by moving the cursor to the vertical line between C3 and C4. Wait for the hollow cross pattern to change into a cross with two horizontal arrows and press the mouse button and drag the arrows to the right to widen Col. 3. After widening the column, put the cursor at the beginning of the title so as to make the disappeared letters reappear. To perform the sorting, go to **Manip > Sort** and, when the dialog box appears, fill the dialog boxes by highlighting and selecting as follows:

---

1) *Sort column(s):* ⌑Count⌑ (note no quotation marks);
2) *Store sorted columns(s) in:* ⌑'Sorted Count'⌑ (note quotes);
3) *Sort by column:* ⌑Count⌑ ☐ Descending (leave box unticked);
4) Click on ⌑*OK*⌑.

Confirm that this gives the numbers of counts for the various occupancy numbers as listed in Col. 3 of Table 6.1. Note that these sorted counts are no longer related to the individual *Fields* in C1.

Table 6.2 Start of MINITAB worksheet for Helber slide counts (from Figure 6.1)

C1 Field	C2 Count	C3 Sorted Count	C1 Field	C2 Count	C3 Sorted Count
1	2	0	3	2	2
1	3	0	3	3	2
1	4	0	3	3	2
1	2	0	3	1	2
1	5	1	3	2	2
1	0	1	3	1	3
1	1	1	3	1	3
1	3	1	3	5	3
1	1	1	4	2	3
1	2	1	4	2	3
1	4	1	4	1	3
1	2	1	4	4	3
1	4	1	4	4	3
1	3	1	4	3	3
1	1	1	4	2	3
1	7	1	4	3	3
2	6	1	4	2	3
2	1	1	4	3	3
2	5	1	4	5	3
2	1	1	4	3	3
2	2	1	4	1	4
2	3	1	4	2	4
2	4	2	4	1	4
2	2	2	4	0	4
2	4	2	5	2	4
2	1	2	5	4	4
2	2	2	5	0	4
2	3	2	5	1	4
2	2	2	5	5	4
2	3	2	5	1	4
2	2	2	5	3	4
2	1	2	5	2	4
3	6	2	5	4	5
3	2	2	5	2	5
3	3	2	5	0	5
3	2	2	5	2	5
3	1	2	5	4	5
3	4	2	5	3	6
3	2	2	5	1	6
3	4	2	5	1	7
continued above right					

6.2.4 Graphic displays

As is customary in the analysis of a fairly large data set, the plotting of a histogram should be an early consideration. Here, it will be useful to ask MINITAB to do the plot with the addition of grid lines so as to facilitate the accurate reading of the counts for each occupancy number, as represented by the bar heights of the histogram. Therefore go to **Graph > Histogram** and, when the dialog box appears:

1) Select and enter under *Graph variable(s):* [Count] (note no quotation marks);

2) Click on *Frame* near the bottom of the window and select *Tick*;

3) When the internal *Frame* window for *Tick* opens, leave the *X* entries with their default settings, but alter the *Y* entries to have 3 under *Number of Major* and 9 under *Number of Minor* (this refers to tick marks). Then close this sub-dialog box;

4) Open *Frame* again but this time select *Grid*.

5) Enter *Y* under *Direction* and leave everything else in the default settings. Close the sub-dialog box;

6) Click on [OK] in the main box.

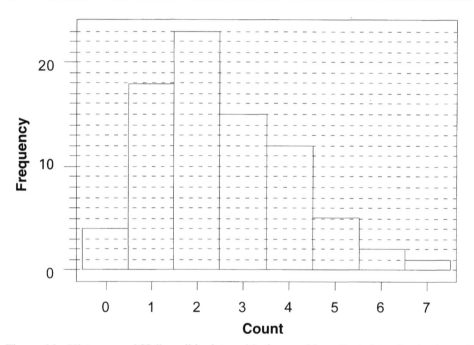

Figure 6.2 Histogram of Helber slide data, with the graphics adjusted to give horizontal lines (see text)

The output should be the histogram in Figure 6.2 that has the same slightly asymmetric shape as the tick diagram in Col. 2 of Table 6.1. The bar heights of the histogram can be read off accurately and seen to correspond to the values for occupancy number totals in Col. 3 of Table 6.1.

Another graphical display for the preliminary inspection of grouped data is the *point plot*, which here can be used to explore possible differences between the different *fields*. Using the procedure of §1.7.1 with *'Jitter'* values of 0.05 for *x* (field) and 0.001 for *y* (count), the graph in Figure 6.3 emerges. It suggests that there are

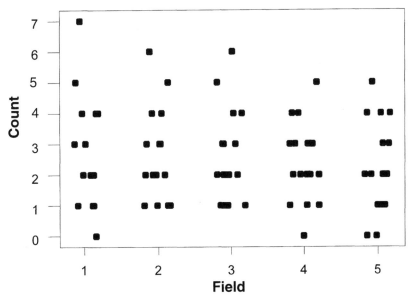

Figure 6.3 Point plot, by *Field* of the Helber count data, with the *Jitter* option used to minimize coincidence of points in a horizontal direction

no major differences either in the average count, or in the spread of counts, in the different *fields*.

6.2.5 Summary statistics

For the further processing of the data, only a selected set of the usual *summary statistics* is needed, rather than the full 'default' output. These can be obtained by entering **Stat > Basic Statistics > Store Basic Statistics** and, when the dialog box appears:

1) Select and enter under *Variables:* $\boxed{\text{Count}}$;
2) Click on *Statistics* in the middle of the box to open the interior menu;
3) From the large number of items offered, tick only *Mean, Sum* and *N-non missing*;
4) Click on $\boxed{OK}$ to close this sub-box and on $\boxed{OK}$ again to close the main box.

The Summary Statistics requested appear in the first three empty columns on the worksheet as:

C4	C5	C6
Mean 1	Sum 1	N1
2.5125	201	80

This reports that the 80 Helber squares yielded a total count of 201 *E. coli* cells, giving a mean count per square of 2.5125. There is no logical inconsistency between each square having only whole numbers of bacterial cells, and yet the average count per square *not* necessarily being a whole number. It is like the average number of children in a family being 2.3, which does not imply that children can exist as fractions of a child. The 2.5125 and 2.3 are *parameters* (constants) that describe the *population* (of Helber squares and human families), whose raw data consist only of whole-number values.

It is also useful to categorize the summary statistics by *Field*. This is achieved in the same way as before, except that after Step 2 above, move the cursor to the *By variables (optional):* box and highlight and select *Field*, the rest of the procedure being as before. This will give the same summary statistics as before, but categorized by *Field*, and appearing in the next set of unoccupied columns on the worksheet as:

C7	C8	C9	C10
By Var 2	Mean 2	Sum 2	N2
1	2.7500	44	16
2	2.6250	42	16
3	2.6250	42	16
4	2.3750	38	16
5	2.1875	35	16

This shows that each field on the Helber slide had $N = 16$ squares, that the total cell count per field ranged from 35 to 44, and that the mean counts per square varied from 2.1875 to 2.7500 correspondingly. In §6.6.1 below, the chi-square test is used to explore whether these differences in total counts per field are statistically significant.

6.3 POISSON DISTRIBUTION FOR INTERPRETING THE HELBER DATA

6.3.1 Theoretical Poisson

With normally-distributed measurement data, the mean ($\bar{X}$) and standard deviation (*s*, st. dev.) of the data were calculated to estimate the 'underlying' population parameters μ and σ respectively. With the Poisson distribution there is only *one* defining parameter *m*, which is the mean count per square (or equivalent mean in other contexts such as the daily count of umbrellas left on buses, etc.). The formula for the Poisson distribution is given in the Further Notes (FN 6.1) at the end of this chapter. Meanwhile, the worksheet is continued by entering, at the next available column, a short column of the *occupancy numbers* (*x*). With these in place from 0 to 7, the corresponding Poisson frequencies can be obtained from MINITAB. The first empty column on the worksheet is C11, therefore label it *Occ No* and type in the values from 0 to 7, corresponding to those in the Helber data. It is convenient, for later use, to tabulate the *observed counts* for each occupancy number and this has been done in C12 (abbreviated to *Obs Count* on the MINITAB worksheet) of Table 6.3. Adjacent to this, label C13 with *Pois freq*, to receive the Poisson frequencies from MINITAB. Then go to **Calc > Probability Distributions > Poisson** and, when the dialog box appears:

---

1) Leave •*Probability*, the default option, unaltered;

2) At •*Mean:* insert $\boxed{2.5125}$, the mean count per square on the Helber slide;

3) Opposite •*Input column:* highlight and select $\boxed{\text{'Occ No'}}$;

4) Opposite •*Optional storage:* highlight and select $\boxed{\text{Pois freq}}$;

5) Click on $\boxed{OK}$.

Table 6.3 Poisson distribution with the same mean (m = 2.5125) as the Helber slide data. Against each occupancy number (C11) is shown the Poisson frequency, as a decimal fraction of 1.0 (C13), as percent (C14) and as the expected count over 80 squares (C15). The final column shows the differences between the observed and the expected counts

C11 Occupancy No.(x)	C12 Observed Count	C13 Poisson Frequency	C14 Poisson %	C15 Expected Count (80 squares)	C16 C12–C15
0	4	0.081	8.1	6.5	–2.5
1	18	0.204	20.4	16.3	1.7
2	23	0.255	25.5	20.5	2.5
3	15	0.214	21.4	17.1	–2.1
4	12	0.135	13.5	10.8	1.2
5	5	0.068	6.8	5.4	–0.4
6	2	0.028	2.8	2.3	–0.3
7	1	0.010	1.0	0.8	0.2
Total	80	0.995	99.5	79.7	0.3

Note: the output from MINITAB was rounded to three significant figures.

This gives for each occupancy number, the theoretical Poisson frequency as a decimal fraction of 1.0 (C13 of Table 6.3). Note that the MINITAB output with six figures has been reduced to three or four figures for simplicity in the table. Since it is useful to have the frequencies also expressed in percent (Pois %), the next column (C14) contains the C13 values (before decimal truncation) multiplied by 100. This was calculated on MINITAB by going to **Calc > Calculator** and, when the dialog box appears:

---

1) In *Store result in variable:* insert $\boxed{\text{Pois \%}}$ by highlighting and double-clicking that column in the listing of the columns;

2) MINITAB uses the asterisk symbol (*) for multiplication. Therefore to multiply each Poisson frequency by 100, insert *100** in the *Expression:* box and then highlight and double click on *Pois freq* so that the *Expression:* box now contains $\boxed{\text{100*'Pois freq'}}$ (note the single quotes);

3) Click on $\boxed{OK}$ to activate the calculation.

This gives the Poisson distribution in C14 of Table 6.3 as a series of percentages against the occupancy numbers and with a peak frequency of 20.5% at the occupancy number of 2. By a procedure similar to the above, the values in C13 can also usefully be multiplied by 80 instead of 100 as in C15. This is a column of so-called *expected counts* (over 80 squares), and facilitates the comparison of the *observed counts* in C12, i.e. the actual data, with what would be expected (Exp Count) from a perfect fit to a Poisson distribution with $m = 2.5125$. One measure of the goodness-of-fit is the overall difference between the observed and the expected counts as presented in C16. This subtraction was done as before with the **Calc > Calculator** function but with $\boxed{\text{C12–C15}}$ in Step 1 above, and $\boxed{\text{'Obs Count'–'Exp Count'}}$ in the *Expression* box. Note in Col. 16 that the differences between the actual ('observed') counts and the theoretical ('expected') counts is quite small, and that with the adding and subtraction the net difference is only 0.3. The statistical significance of the differences is explored in the chi-square test in FN 6.3, except that in this test the quantity calculated is the *square* of (Observed count minus Expected count) that is then divided by the Expected count.

The column summations in Table 6.3 can be done on MINITAB, with the results being displayed in the next unoccupied columns from C17 onwards. Go to **Calc > Calculator** and, when the dialog box appears:

---

1) In *Store result in variable:* insert $\boxed{\text{C17}}$ by typing it in;

2) Go to *Functions* and scroll down the menu until *Sum* appears. Highlight it and double-click so that it is transferred to the *Expression* space, where it will appear as SUM(number);

3) Delete the *number* in *SUM(number)* and replace it with *Pois freq* by highlighting and double-clicking this column from the list. The *Expression* space should now show $\boxed{\text{SUM('Pois freq')}}$;

4) Click on $\boxed{OK}$ to activate the calculation.

Proceed in a similar way to total the other columns and note that the frequency values delivered are, within rounding limitations, equal respectively to 1.0, 100 and 80. Having noted down these values, the worksheet columns just used can be cleared with the **Erase variable** command in the **Manip** window.

6.3.2   Graphical plot of the Poisson distribution

Figure 6.4 (top) is a graphical plot of the Poisson distribution for $m = 2.5125$. It was done initially as a point plot on the Plot menu and the vertical bars added by opening the toolbar with the right-hand button on the mouse and choosing the line symbol. It is customary to represent the Poisson distribution as a set of unconnected vertical lines in order to emphasize its discontinuous nature. Thus the points of the Poisson should not be joined up to each other, nor should a curve be fitted around the top outline (but see below!), since this would imply the existence of values in-between the occupancy numbers.

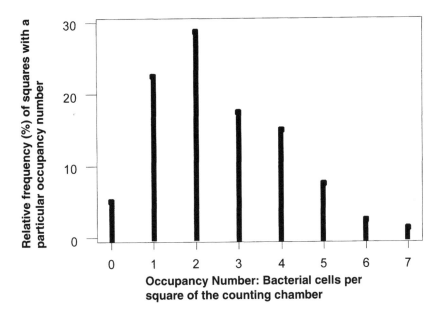

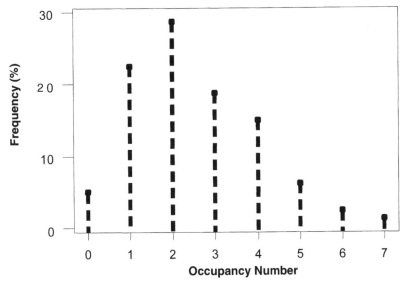

Figure 6.4 Frequency plots. Top, Poisson distribution for *m* = 2.5125; bottom, Helber slide count data for comparison

The corresponding point plot, with added vertical lines (dotted for contrast with the Poisson), for the Helber slide data is in Figure 6.4 (bottom). It was produced by first creating an extra column for the Helber data expressed as frequency %. This was done by multiplying the *observed counts* in C12 of Table 6.3 by 1.25 so as to express them as a percentage (i.e. 100/80 = 1.25, the factor needed). Comparison of the two graphs in Figure 6.4 suggests that there is a close resemblance of the Helber data to a Poisson distribution with the same mean.

6.4 CHANGEABLE SHAPE OF THE POISSON DISTRIBUTION

6.4.1 Poisson distribution for *m* = 1.0

To illustrate this section, suppose that the number of bacterial cells counted over the 80 squares of the Helber slide was exactly 80, thus giving a mean count of 1.000 per square. Some folk, when presented with this proposal, think that each Helber square will therefore contain one bacterial cell. But this is not so, because of the randomness of the distribution of the bacteria. Therefore some of the squares will be empty ($x = 0$) and some squares will have two or more cells in compensation, to give the overall average of 1.0. Application of the Poisson formula allows prediction of the frequency of empty squares, squares with one cell, two cells and so on. The procedure is as in §6.3.1 with the occupancy numbers of C11 (Table 6.3) above and yields the frequencies set out in C18 of Table 6.4. Two extra manipulations have been performed with the **Calc > Calculator** function as above to give C19 with the frequencies as a percentage, and C20 with the percentages rounded to two decimal places. First label some destination columns on the worksheet:

C18: with *P for m =1* (meaning *Poisson probabilities for m = 1*);
C19 with *P%m=1* (i.e. *Poisson probabilities for m = 1, expressed as percent*);
C20 with *P%m=1R* (shorthand for *Poisson frequencies for m = 1, expressed as percent and rounded to two decimal places*.

The columns for the Poisson probabilities as a decimal fraction of 1.0 and as a percentage are obtained as above. To obtain the rounding function, go to **Calc > Calculator** and, when the dialog box appears:

---

1) In *Store result in variable:* insert $\boxed{\text{'P\%m=1R'}}$ by highlighting in the column menu and double-clicking;

2) Go to *Functions* and scroll down the menu until *Round* appears. Highlight it and double-click so that it is transferred to the *Expression* space, where it will appear as ROUND(number, num_digits);

3) Delete the *number* in *ROUND(number)* and replace it with *'P%m=1'* by highlighting and double-clicking this column from the list. Then delete *num_digits* and replace it with 2 (the number of decimal places wanted); the *Expression* space should now show $\boxed{\text{ROUND('P\%m=1',2)}}$;

4) Click on $\boxed{OK}$ to activate the calculation.

This should give the figures in Col. 20 of Table 6.4. Note that the rounding by MINITAB is not done simply by chopping off the decimal places beyond 2, but by adjusting the second decimal place upwards if required for best representation of the removed figures. Table 6.4 and the frequency plot in Figure 6.5 (top) show that

Table 6.4 MINITAB worksheet for the calculation of frequencies for a Poisson distribution with $m = 1.0$ (continuation of Table 6.3)

C11 Occupancy No. (x)	C18 P for $m=1$	C19 P%$m=1$	C20 P%$m=1$R
0	0.367879	36.7879	36.79
1	0.367879	36.7879	36.79
2	0.183940	18.3940	18.39
3	0.061313	6.1313	6.13
4	0.015328	1.5328	1.53
5	0.003066	0.3066	0.31
6	0.000511	0.0511	0.05
7	0.000073	0.0073	0.01

the Poisson distribution for $m = 1.0$ has the first two bars of exactly equal height at a frequency close to 37%. This means that a Helber slide with an average of one bacterial cell per square would have (within random-sampling fluctuations) 37% of the squares empty, another 37% with one cell, 18% with two cells, and rapidly diminishing percentages for the higher occupancy numbers. Likewise with other Poisson situations with $m = 1$ such as umbrellas left on buses, an average of one lost umbrella per day (say, during the past year of record keeping) would raise the expectation that on about 37% of the days, no umbrellas ($x = 0$) would have suffered this fate.

6.4.2 Poisson distribution for $m = 10$

Repeating the above procedures, but with $m = 10$, results in a dramatic change in shape for the Poisson distribution (Figure 6.5, bottom) whose outline now resembles a normal distribution curve, except that the Poisson still differs from the normal in being *discontinuous*. For MINITAB to deliver this graph requires a longer set of *occupancy numbers*, for which the set from 1 to 20 was entered, by the *Make Patterned Data* command, in the next available column (C21) on the worksheet (Table 6.5). This column was labelled *Occ No 2*, meaning the No 2 set of occupancy numbers. Adjacent to it are C22, headed *P for m=10*, to receive the Poisson frequencies for m = 10, and C23 with *RP m=10*, for the *rounded frequencies for a Poisson distribution for m = 10*. With these labelled columns in place, the MINITAB procedure followed that used for $m = 1$. Thus for rounding to two decimal places the frequencies in C22, the entry for the **Calculator** *Expression* space was *ROUND(100*'P for m=10', 2)*.

6.4.3 Approximation of Poisson to a normal distribution

An important feature of the Poisson distribution is that as m becomes what the statisticians call 'large', meaning about 10 or greater, the Poisson distribution starts to approximate quite closely to a normal distribution. As m increases still further, the closeness of this approximation steadily improves, so that for $m = 20$ or more, the Poisson can for practical purposes be treated as being normal. At this point the fact that it is discontinuous is set aside. And under these circumstances the mean

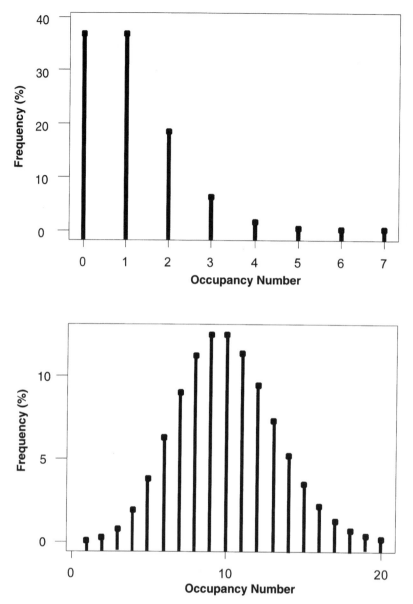

Figure 6.5   Frequency plots of the Poisson distribution for: top, $m = 1.0$; bottom, $m = 10$

(m) of the Poisson distribution is taken as being the same as the population mean (μ) of the normal distribution to which it approximates.

What about the standard deviation of the Poisson? The answer is that when the Poisson distribution of mean (m) approximates to a normal distribution, its standard deviation (σ) is equal to the *square root of m*. Thus if the mean count m in the Poisson is, say, 25, and this is taken as approximating to a normal distribution of $\mu = 25$, the st. dev., σ, is $\sqrt{25} = 5$.

Table 6.5 Continued MINITAB worksheet for getting the frequencies for the Poisson distribution with $m = 10$

C21 Occ No 2	C22 P for $m=10$	C23 RP $m = 10$
Occupancy No.	Frequency	Frequency (%)
1	0.000454	0.05
2	0.002270	0.23
3	0.007567	0.76
4	0.018917	1.89
5	0.037833	3.78
6	0.063055	6.31
7	0.090079	9.01
8	0.112599	11.26
9	0.125110	12.51
10	0.125110	12.51
11	0.113736	11.37
12	0.094780	9.48
13	0.072908	7.29
14	0.052077	5.21
15	0.034718	3.47
16	0.021699	2.17
17	0.012764	1.28
18	0.007091	0.71
19	0.003732	0.37
20	0.001866	0.19

With the Helber data, the mean count per square of 2.5125 is too low to be taken as approximating to a normal distribution. Also the plot (Figure 6.5) is too asymmetrical to be taken as being normal. However, the data can be regarded in an alternative fashion. Over the 80 Helber squares scanned, a total of 201 bacterial cells were counted. So instead of taking the Poisson m as 2.5125 per small square, the $m = 201$ is equally valid as a count of the large area equal to 80 small squares. A Poisson mean as large as 201 amply fulfils the requirements of being a very close approximation to the normal $\mu = 201$. Therefore the st. dev. of this count is $\sqrt{201} = 14.18$. This can then be used to calculate the SE mean and the 95% confidence intervals of the count.

6.5 STANDARD ERROR AND CONFIDENCE INTERVALS OF COUNTS

6.5.1 Large counts

For the mean counts of randomly-distributed objects or events (per unit of area, volume or time, as appropriate) of 20 or more, the Poisson distribution can be taken as approximating to the normal. So to continue with the Helber data, the mean count of 201 cells over 80 squares, with its st. dev. of $\sqrt{201} = 14.18$, also has a standard error of the mean of 14.18. In other words, the number 201 is taken as

Table 6.6 Relative width of confidence intervals (CIs) of counts between 100 and 100 000

Count	St. dev.	95% CI	Spread of CIs around count	Spread of CIs as % of count
100	10	80, 120	±20	±20
1000	31.6	938, 1062	±62	±6
10 000	100	9804, 10 196	±196	±2
100 000	316	99 380, 100 620	±620	±0.6

being a sample of $N = 1$ observations from a normal distribution of $\mu = 201$ and $\sigma = 14.18$. Thus the SE mean = 14.1.

To obtain the 95% confidence intervals, the SE mean is multiplied by 1.96, which technically is the value of the t-statistic for infinity degrees of freedom. Thus the 95% CIs are $201 \pm (1.96 \times 14.18) = 201 \pm 27.8 = 173$ and 229. The 201, 173 and 229 can then be divided by 80 to express the final result in terms of the small square as the unit of counting. Therefore the count per small square rounded to two decimal places, with 95% CIs is 2.51 (95% CIs 2.16 and 2.86). Remembering (§6.2.1) that the volume of fluid above each small square is 0.5×10^{-7} cm^3, the mean count of bacterial cells per ml of suspension is 5.0×10^7, with 95% confidence intervals of 4.3 and 5.7×10^7. Note how the number of decimal places is finally reduced to a reasonable value (in relation to the width of the confidence intervals and the uncertainty so implied), but only at the very end of the calculation.

With very large counts, such as are commonly obtained in radioactivity experiments or Coulter counts of cell populations, the confidence intervals can diminish to very narrow ranges, as the counts become very large. Table 6.6 illustrates this with counts between 100 and 100 000. With a count of 10 000, the spread of confidence intervals around the mean is $\pm 2\%$, while with 100 000 counts the spread is reduced to $\pm 0.6\%$.

6.5.2 Small counts

With counts of less than about 20 it is possible to calculate 95% confidence intervals, but they become very wide as the count becomes small. Suppose for example with the Helber slide that there were only two bacterial cells over the whole 80 squares. The normal approximation would not apply but nevertheless it is possible to calculate confidence intervals of the two cells counted. A more elaborate formula has to be used and it is convenient to obtain the CIs of small counts from a statistical table. Table 6.7 presents the 95% CIs for counts between 1 and 20, and can be consulted in these circumstances. It will be seen that a count of 2, for example, has 95% CIs of 0.24 and 7.2 which of course are so hopelessly wide as to be of little use (except to show how quantitatively poor the information is in such a low count). Note, however, that a count of 20 has 95% CIs of 12 and 31 These values are close to the values yielded by the formulae used above, since $\sqrt{20} = 4.472$, which when multiplied by 1.96 gives CIs of $20 \pm 8.76 = 11.2$ and 28.8. However, from the practical standpoint of obtaining reliable results, it would be much better to have a larger count in the first place instead of trying to squeeze accurate information out of counts that are intrinsically too small.

Table 6.7 Ninety-five percent confidence intervals of counts between 1 and 20 (rounded to 2 significant figures)

Count	95% CI	Count	95% CI	Count	95% CI	Count	95% CI
1	0.025, 5.6	6	2.2, 13	11	5.5, 20	16	9.1, 26
2	0.24, 7.2	7	2.8, 14	12	6.2, 21	17	9.9, 27
3	0.62, 8.8	8	3.4, 16	13	6.9, 22	18	11, 28
4	1.1, 10	9	4.1, 17	14	7.6, 23	19	11, 30
5	1.6, 12	10	4.8, 18	15	8.4, 25	20	12, 31

6.6 ANALYSIS OF DIFFERENCES BETWEEN COUNTS BY CHI-SQUARE

6.6.1 Application of chi-square

To analyse a set of counts for possible statistical differences, the chi-square test may be applied. Chi-square is also called chi-*squared* and may be represented with its Greek letter as χ^2. Its procedure follows the same general pattern as other tests of significance, such as the *t*-test or ANOVA. With MINITAB, however, there is a problem, because although it has considerable chi-square capability, there is a gap that does not allow the direct input, into any of the pull-down menus, of sets of counts such as the five *Fields* (Figure 6.1) in the Helber slide. So in what follows, MINITAB is applied where possible and is supplemented, as necessary, by conventional statistical working with a formula and tables. Further insights and applications of chi-square are provided in the Further Notes (FN 6.2 and FN 6.3).

Referring back to §6.2.5, the five *Fields* of the Helber slide yielded counts of 44, 42, 42, 38 and 35 bacterial cells. The question now asked is whether this amount of variation is *within the limits of random-sampling fluctuations* or whether the differences between *Fields* are *real*. For example, *real* differences might exist if the surface of the slide was uneven so that some areas were deeper than others, and therefore held a larger volume, with a correspondingly larger count of bacterial cells. A similar question might be asked of five culture plates, or five weeks of leaving umbrellas on buses. The formula for chi-square (χ^2) is:

$$\chi^2 = \Sigma \, (O - E)^2 / E \qquad \text{Eq. 6.1}$$

where O is the *observed* counts 44, 42, 42, 38 and 35, and E is the *expected* count (*expected* if there was no variation beyond sampling fluctuations). The expected count is the arithmetic average of the observed counts and is 40.2. The fact that E is not a whole number is immaterial since it is an abstract quantity and not the count of an actual *Field*. It is like the average 2.3 children in a family. Capital *sigma* (Σ) is the summation sign, so working the equation with a pocket calculator gives:

$$\chi^2 = [(44 - 40.2)^2 + (42 - 40.2)^2 + (42 - 40.2)^2 + (38 - 40.2)^2 + (35 - 40.2)^2] \div 40.2$$
$$= [(3.8)^2 + (1.8)^2 + (1.8)^2 + (-2.2)^2 + (-5.2)^2]/ \, 40.2 \text{ (note how the minus signs}$$
disappear on squaring)
$$= [14.44 + 3.24 + 3.24 + 4.84 + 27.04]/40.2$$
$$= 52.8/40.2 \, = \, 1.31$$

This final result of $\chi^2 = 1.31$ is the *found* value of the chi-square statistic for $(N - 1) = 4$ degrees of freedom (the degrees of freedom being one less than the

Table 6.8 Continued MINITAB worksheet for calculating chi-square for analysis of differences between Helber slide *Fields* and for manipulated data

C24 Field Obs	C25 Field Exp	C26 Chi-Square	C27 Field Obs2	C28 Chi-Square2
44	40.2	1.31343	54	10.7662
42	40.2		42	
42	40.2		38	
38	40.2		38	
35	40.2		25	

C24: *Field Obs* = 'Observed' numbers of bacterial cells in each Field;
C25: *Field Exp* = 'Expected' numbers of bacterial cells in each Field;
C26 *Chi-Square*, calculated from C24 and C25;
C27: *Field Obs2* = as C24, but 10 added and subtracted from first and last observation respectively;
C28: *Chi-Square2* = calculated from C25 and C27.

number of groups of counts). This point can be reached with MINITAB by entering the observed and expected field counts on the worksheet in the next empty columns (Table 6.8) and then entering SUM(((C24-C25)**2)/C25) in the *Expression* space of the **Calc > Calculator** dialog box (Figure 6.6). The chi-square result 1.31343 then appears in the specified destination column C26 (Table 6.8).

As with the *t*-test, there is a null hypothesis that states 'the extent of scatter of the Field counts is no more than would be expected by random-sampling fluctuations'.

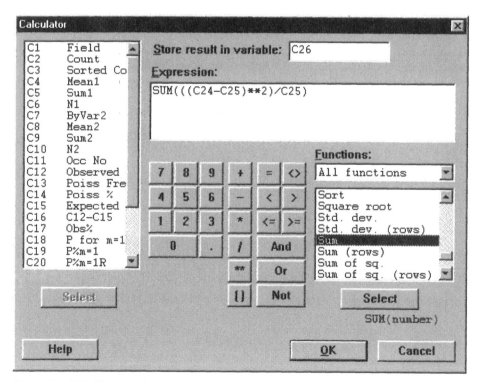

Figure 6.6 MINITAB *Calculator* dialog box for calculating chi-square for the Helber *Field* results

Table 6.9 Significance points for chi-square

Degrees of freedom	Value of chi-square for probability (%)		
	5	1	0.1
1	3.84	6.63	10.8
2	5.99	9.21	13.8
3	7.81	11.3	16.3
4	9.49	13.3	18.5
5	11.1	15.1	20.5
6	12.6	16.8	22.5
7	14.1	18.5	24.3
8	15.5	20.1	26.1
9	16.9	21.7	27.9
10	18.3	23.2	29.6

To test this hypothesis, the tabulated probability (P) value for chi-square with 4 degrees of freedom is needed. This can be obtained from MINITAB as described in FN 6.2, but for convenience here is presented in the chi-square table (Table 6.9) for degrees of freedom up to 10. The critical probability value (P) for statistical significance (i.e. rejecting the null hypothesis at the $P = 5\%$ level) has a chi-square of 9.49 for 4 degrees of freedom. Therefore the found chi-square of 1.31 is well within the range expected from sampling fluctuations and the observed scatter of *Field* counts does not represent significant variation.

6.6.2 Chi-square from MINITAB

To obtain the probability value for chi-square directly from MINITAB, rather than from Table 6.9, first type in to an empty column on the worksheet (say C30), the found value of chi-square whose *P*-value is required, and identify the adjacent column (say, C31) as the destination for the result. Suppose that 9.49 is put into C30 as the value of chi-square whose probability is wanted. Then go to **Calc > Probability Distributions > Chi-Square**, and when the dialog box appears:

1) Select *Cumulative probability*;
2) Enter the *Degrees of freedom* as $\boxed{4}$ (or whatever the number may be in other examples);
3) *Input column*: $\boxed{C30}$;
4) *Optional storage*: $\boxed{C31}$;
5) $\boxed{OK}$.

This will deliver the value 0.95 into C31. To obtain the conventional *P*-value, subtract the C31 value from 1.0 to get 0.05, the tabulated 5% point of chi-square for 4 DF and $P = 5\%$ (Table 6.9).

6.6.3 Manipulated *Field* counts

As an exercise to see how much extra scatter is required to make the *Field* counts significantly different, they will now be manipulated so as to make the highest one somewhat higher and the lowest one a good bit lower. This will considerably increase the extent of scatter. By doing the manipulation symmetrically in this way the *Expected* count is not affected. Thus if the count of 44 is increased by 10, to 54, and the count of 35 is lowered by 10 to 25, there is then a set of manipulated *Field* counts of 54, 42, 42, 38 and 25. These manulations can be done in the next free column (C27) of Table 6.8, which can be labelled *Field Obs2*. The destination column (C28) for the new chi-square is *Chi Sq 2* (Table 6.8). Entering SUM((((C24-C27)**2)/C25) in the *Expression* box of the **Calc > Calculator** dialog box gives a *Found* Chi-square = 10.7662. This exceeds the value 9.49 for 4 degrees of freedom in the chi-square table (Table 6.9) and leads to the *rejection* of the null hypothesis at the $P = 5\%$ level. Thus the manipulated extent of scatter in C27 is no longer reasonably explained by sampling fluctuations, and the difference in counts between *Fields* is declared to be 'statistically significant' or 'real'.

6.6.4 Important restriction on use of chi-square

In the above examples, the count figures being put into the chi-square formula were in the order of 30–50. If they had been below about 5, the chi-square test would no longer be applicable. Thus chi-square can not be used to explore in the same way the scatter of the counts, averaging around 2.5, in the individual small squares of the Helber slide. This same restriction would equally apply to other types of objects or events being counted. For further advice, an expert statistician should be consulted.

The restriction applies actually to the *Expected* count, none of which should be less than about 5. So that if there were two counts of 4 and 12 to be tested for heterogeneity by chi-square, the expected count would be $(4 + 12)/2 = 8$, and the chi-square test could be performed. It would yield:

$$\chi^2 = (4-8)^2 /8 + (12-8)^2 /8 = 2 + 2 = 4$$

which, with *one* degree of freedom, would be just significant (Table 6.9) at the $P = 5\%$ level. On the other hand, two counts of 2 and 6 could not be tested by chi-square because the expected count would be only 4.

6.7 APPLICATION OF ANOVA

6.7.1 Restrictions on ANOVA

As emphasized in an earlier chapter, the ANOVA procedure is intended for data that meet three requirements:

- Obtained by a method involving random allocation of subjects to experimental groups;
- Involving a *continuous* variable that also is (at least approximately) normally distributed;
- Having within-group st. devs that do not differ significantly.

With count data, appropriate design can satisfy the first requirement, but the other criteria present problems that require attention. Data such as the small-number counts from the Helber slide fail to meet the *normality* criterion both by being discontinuous and by coming from a distribution that is asymmetrical. So although it is physically possible to do one-way Anova with MINITAB, to analyse for possible differences between *Fields* on the Helber slide, it would be a misuse for the reasons indicated. However, it was appropriate to analyse for possible between-*Field* differences by chi-square applied to the individual *Field* totals.

However, all is not lost and since ANOVA is such a powerful tool, the search for solutions may be worth pursuing. An opportunity appears if the data consist of counts that are about 10 or more, because from this region onwards, the Poisson distribution becomes an ever closer approximation to a normal distribution, albeit still discontinuous.

The more serious difficulty of *inhomogeneity of variances* then has to be addressed. This follows from the feature of the Poisson that the *variance* is equal to the *count* and the st. dev. is therefore equal to the *square root of the count*. Therefore as the count gets larger, so do its variance and st. dev. also increase. For example, a group of experimental counts averaging around nine would have a st. dev. of about 3, while a group with an average count of 100 would have a st. dev. near to 10. (The st. devs would not be the exact square root because of the deviation from normality of a count of 9 and also sampling fluctuations.) Such a large disparity of st. devs would invalidate the ANOVA, which requires that there be no significant differences between the st. devs of the different groups. However, this problem of *heteroscedasticity*, to use the technical name, is solved by a square-root transformation of the count data, as the following example will illustrate.

6.7.2 Constructing an illustrative example

This illustrative example is an artificial data set consisting of four groups, each containing five random counts, with group means of approximately 9, 25, 81 and 121. These numbers were chosen so as to have corresponding st. devs of about 3, 5, 9 and 11. To generate the data, a new worksheet, No. 2, was opened in the *Chap 6 Helber* Project. This is not because Worksheet No. 1 was running out of space, but because there was no special virtue in adding further columns on the right, with all the scrolling involved, especially for a new set of data that would be unrelated. Figure 6.7 shows the appearance of Worksheet No. 2 after completion. Since most of the operations have already been described for Worksheet No. 1, the following is only a summary outline:

Col. 1 *Group*: was filled with five each of the numbers 1–4 with the *Make Patterned Data* command;

Col. 2 *Count*: was filled at the end, by a *Stack* command applied to C3, C4, C6 and C8 after they had been obtained;

Col. 3 *Mean 9*: contains five randomly-generated counts from a Poisson distribution with mean = 9;

Col. 4 *Mean 25*: as C3, but from a Poisson with mean = 25;

Col. 5 *Mean 81*: contains five randomly-generated values from a normal distribution of mean = 81 and st. dev. = 9. Ideally the Poisson function would have been used as in C3 and C4, but MINITAB does not give random Poisson counts for numbers above 49;

Col. 6 *Mean 81R*: contains the values in C5 *rounded* to whole numbers with the

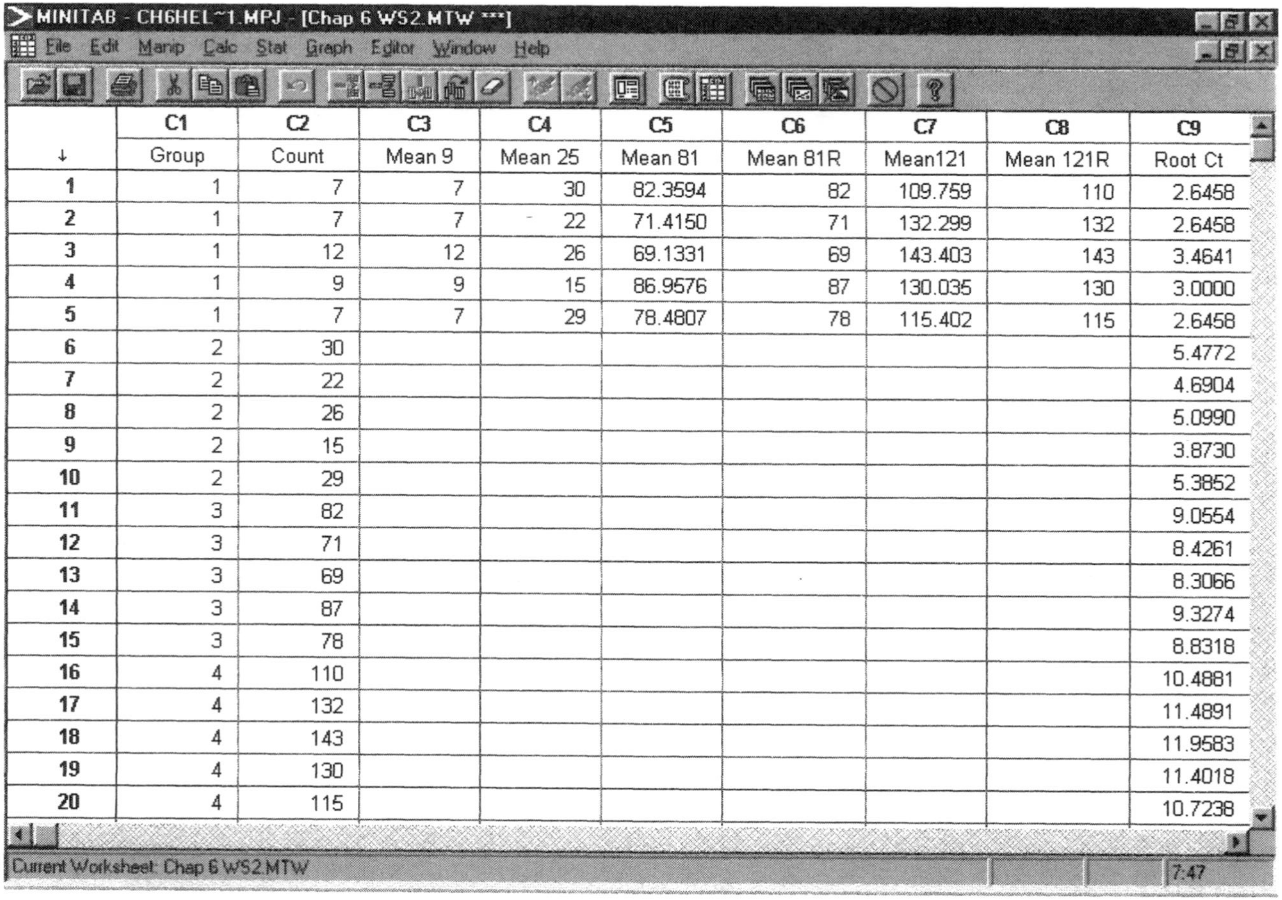

	C1	C2	C3	C4	C5	C6	C7	C8	C9
	Group	Count	Mean 9	Mean 25	Mean 81	Mean 81R	Mean121	Mean 121R	Root Ct
1	1	7	7	30	82.3594	82	109.759	110	2.6458
2	1	7	7	22	71.4150	71	132.299	132	2.6458
3	1	12	12	26	69.1331	69	143.403	143	3.4641
4	1	9	9	15	86.9576	87	130.035	130	3.0000
5	1	7	7	29	78.4807	78	115.402	115	2.6458
6	2	30							5.4772
7	2	22							4.6904
8	2	26							5.0990
9	2	15							3.8730
10	2	29							5.3852
11	3	82							9.0554
12	3	71							8.4261
13	3	69							8.3066
14	3	87							9.3274
15	3	78							8.8318
16	4	110							10.4881
17	4	132							11.4891
18	4	143							11.9583
19	4	130							11.4018
20	4	115							10.7238

Figure 6.7 MINITAB worksheet for the calculation to show the benefit of the square-root transformation of count data (see text)

Round function in the *Calculator* menu and inserting 0 for the number of decimal places wanted. Note how the values in C5 have been rounded to the nearest whole number and not simply had the decimals removed;

Col. 7 *Mean 121*: is as C5 but with 121 and 11 inserted as the parameters of the normal distribution;

Col. 8 *Mean 121R*: contains the rounded values from C7.

With the individual groups now available, the *Stack* command was applied to assemble C3, C4, C6 and C8 as a single column in C2, with the corresponding Group numbers in C1. Finally, the counts in C2 were converted into their square roots (C9) with the *Sqrt* function in the *Calculator* menu. Having generated this instructional set of counts, the benefit of the square-root transformation will become apparent.

6.7.3 Benefit of square-root transformation

Figure 6.8 (top) is a point plot of the data in C1 and C2 of the new worksheet. It illustrates the increasing scatter (heteroscedasticity) of groups of random-sample counts as the average count increases. Figure 6.8 (bottom) shows how the problem of heteroscedasticity can be overcome by transformation of the counts to their square roots (C9). Confirmation that the square-root transformation of the counts has reduced the differences between the st. devs is provided by the descriptive statistics in Table 6.10. Col. 7 from the left shows that with the untransformed counts, the st. devs ranged from 2.19 to 13.40, a sixfold difference. Whereas with the square roots of the counts, the st. devs went from 0.361 to 0.598, which is less than a twofold variation.

Table 6.10 Mean counts and st. devs without and with the square-root transformation (from Figure 6.7)

Group No.	Mean of the counts		Mean of the square roots of the counts		St. dev. of the counts		St. dev. of square root of the counts
	(Exptd)	(Obs)	(Exptd)	(Obs)	(Exptd)	(Obs)	(Obs)
1	9	8.40	3.0	2.88	3.0	2.19	0.361
2	25	24.40	5.0	4.90	5.0	6.11	0.653
3	81	77.40	9.0	8.79	9.0	7.50	0.426
4	121	126.00	11.0	11.2	11.0	13.40	0.598

Confirmation that the variances (and therefore the st. devs), before the square-root transformation, are significantly different may be obtained by the *test for homogeneity of variances*, used in the last chapter. To do this, go to **Stat > Anova > Homogeneity of variance** and, when the dialog box appears:

1) In *Response:* insert Count by highlighting in the column menu and double-clicking;

2) In *Factors:* insert Group ;

3) Leave *Confidence level:* with its default entry of 95% ;

4) Click on OK to activate the calculation.

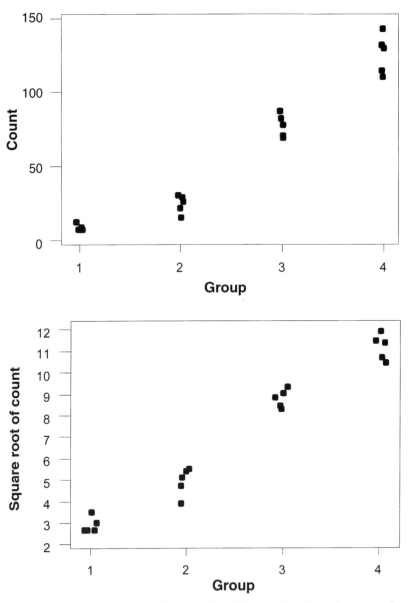

Figure 6.8 Point plots of constructed count data. Top, to show how the scatter increases with the count; bottom, to show how the square-root transformation equalizes the extent of scatter

The output from the tests in Figure 6.9 has the combination of pictorial presentation and test statistics. The Bartlett test statistic for the untransformed data (Figure 6.9, top) had a *P*-value of 0.026, indicating significant heterogeneity of variances. The detail of this heterogeneity is on the left side which presents the 95% confidence intervals for the st. dev. (= sigma = σ) of each group. It shows the progressively increasing st. dev. between Groups 1 and 4 and how the central

Homogeneity of variance test for count

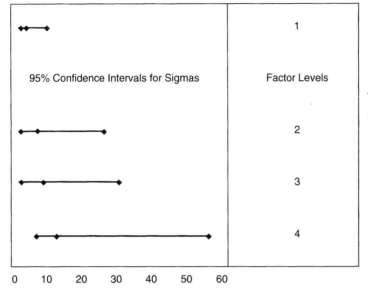

Batlett's Test

Test Statistic: 9.257
P-Value: 0.026

Homogeneity of variance test for square root of count

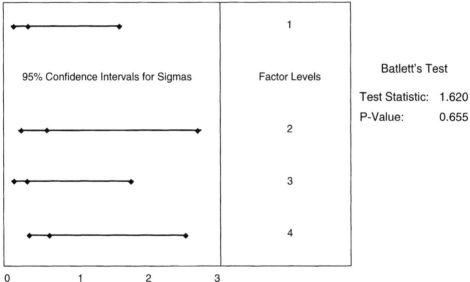

Batlett's Test

Test Statistic: 1.620
P-Value: 0.655

Figure 6.9 Test for homogeneity of variances. Top, count data before square-root transformation; bottom, after transformation

estimates of st. dev. (central point on each line) fall outside the confidence intervals of some other groups. For example, the st. dev. of Group 1 lies outside the CIs of all the other groups. Likewise although the central estimate of Group 2 is within the CIs of Group 3 and also of Group 1, it is outside the CIs of Group 4.

After the square-root transformation of the counts (Figure 9, bottom), the test no longer indicates significant heterogeneity of variances, the *P*-value being 0.655. It is therefore legitimate to perform ANOVA on the square roots of the counts. When this is done, as described previously, the final *P*-value is 0.000, indicating, as would be expected, highly significant differences in the mean counts of the four groups.

Transformations other than square root, which may be worth considering with other count data, are the $\log_{10}$ and the fractional-power transformations. Expert advice should be sought.

6.8   WHERE POISSON FAILS TO RULE

While there are some well-established types of count data to which the Poisson distribution applies (Helber slide, Coulter Counter, radioactivity), there are other kinds of count data to which it definitely does not. Therefore with an unknown system, it is worthwhile examining the nature of the distribution (FN 6.3) to see if the assumption of the Poisson is justified. Departures from Poisson may arise from lack of:

- Independence, i.e. lack of randomness in the distribution of the objects or events in space or time;
- Uniformity of capture units, i.e. the units of space or time in which the individual objects or events are recorded. For example, if the Helber slide contained areas that were shallower or deeper than others, this would influence the number of bacteria appearing above the small squares and hence introduce heterogeneity. Important inputs to this category are the volumetric errors in sample-dispensing, e.g. in pipetting supposedly equal volumes of bacterial suspensions on to culture plates, or radioactive samples into scintillation vials. Thus the actual radioactivity counting might be carried out to a Poisson accuracy (SEM) of $\pm 1\%$ by recording a sufficiently large number of disintegrations (say, 100 000). But if the sample itself has a pipetting error of $\pm 3\%$, then replicate samples might not deliver the consistency of counts expected from the approximately 100 000 disintegrations counted from each.

6.8.1   Aggregation, clustering, contagion and general patchiness

Although it is rash to make sweeping generalizations in biology, I am now going to make one: organisms, macro and micro, *when living in their natural habitats* are commonly *not* Poisson distributed. Instead they tend to have an aggregated, or clustered, type of distribution, rather than being spread around at random (Taylor, 1961). In fact, the richness of the English language in special collective nouns, such as herd, pride, flock, pack, gaggle, shoal, etc., is evidence of the tendency of animals *not* to occur in nature as individuals that are randomly dispersed over the Earth's surface and free from interactions with each other. The same is true of microorganisms colonizing surfaces in the soil or in aquatic habitats and where micro-colonies are commonly observed. Likewise with plants, although seeds and spores may disperse over long distances at random, once they have come to earth and germinated, there is a tendency for further individuals to arise in close proximity to the original mother plant in the habitat to form a cluster. The word 'contagion' in the title refers to the clustered distribution seen in the spread of a contagious disease in a community. Rather like the seeds of a plant arriving from afar to colonize a new habitat, so an infectious agent sets up primary infection in one individual (e.g. animal, person or fruit tree) from which it may spread to other individuals in the vicinity, thereby forming a cluster of infected individuals and giving rise to a 'contagious distribution'. This, again, is a departure from the Poisson distribution, which requires the individuals in the population to be independent and non-interacting with each other.

6.8.2 Under-distribution and over-distribution

As emphasized already, the Poisson distribution requires that the particles, objects or events be *randomly* distributed. There are two main forms of departure, known as *under*-distribution and *over*-distribution, the latter being more common with living organisms in their natural habitats.

Under-distribution is seen, for example, in the arrangement of atoms in a crystal lattice, or of apple trees in regular spacing in an orchard as viewed from an aircraft. With free-living organisms, a state of under-distribution would imply some kind of territorialism leading to a relatively constant spacing of individuals in the environment. The use of '*under*' is because the extent of dispersion is less than what it would be if the distribution were random.

Over-distribution, by contrast, is the opposite phenomenon, of aggregation, patchiness, contagion, etc.

With living things in their natural habitats there may well be both over- and under-distribution existing simultaneously. For example, with fish that occur in shoals, there may be over-distribution if the equivalent of the Helber grid square is 1 km^2 of sea bed. However, in the microhabitat of the shoal itself and taking a unit sampling volume of 1 m^3, there may be under-distribution, since each fish may maintain a fairly constant distance from its neighbours.

6.8.3 Over- and under-distribution of total cell counts

It is convenient to illustrate under- and over-distribution by modifying the Helber slide results. This has been done in Table 6.11 and plotted in Figure 6.10, which contain these 'deviant' distributions, as well as the actual Helber data and the corresponding 'pure' Poisson distribution. The bottom lines of the table show that all four distributions are based on 80 squares with a total of 201 bacterial cells, but with different patterns of occupancy. Thus the under-distribution is unduly peaked compared with the actual Helber or the pure Poisson, while the over-distribution has a large proportion of empty squares but accompanied by a peak of clumping at an *x*-value of 5. With such distributions that depart significantly from the Poisson, it would be wise to seek expert statistical advice so as to avoid the inappropriate use

Table 6.11   Four sets of Helber slide data to illustrate *under-*, *over-*, *actual* and *pure Poisson* distributions. All four sets have 201 bacterial cells counted over 80 squares, giving an average count per square of $m = 2.5125$

| Occupancy number of square | Number of squares containing *x* bacterial cells | | | |
	Under- distribution	Actual distribution	Poisson distribution	Over- distribution
0	10	4	6.49	31
1	5	18	16.30	3
2	12	23	20.47	6
3	41	15	17.14	8
4	11	12	10.79	10
5	1	5	5.41	13
6	0	2	2.27	6
7	0	1	1.13	3
Totals				
No. Squares	80	80	80.0	80
No. Cells	201	201	200.4	201

of Poisson-dependent methods. Non-parametric statistical methods (*q.v.*) may prove useful as a first line of approach.

6.9 SUMMARY

This chapter has used a single set of bacterial cell counts, done with a Helber-slide counting chamber, to illustrate the statistical thinking and approaches that have wide application to many other types of count data. After entry of the counts on to a MINITAB worksheet, the analysis started with graphical displays and continued with summary statistics, including confidence intervals. The Poisson distribution was introduced as the underlying model, and its relation to a normal distribution emphasized. The application of chi-square to the analysis of count data was illustrated, and the benefit of the square-root transformation for analysis of variance of counts was demonstrated. Finally, it was pointed out that living organisms in their natural environments are rarely in a Poisson distribution, but tend to show aggregation or clustering.

FURTHER NOTES

FN 6.1 Formula for the Poisson distribution

The equation for the Poisson distribution is:

$$P_x = e^{-m} . m^x/x! \qquad \text{Eq. 6.2}$$

where P_x is the probability, or frequency, of finding (in Helber terms) squares with *x* bacterial cells in them, *x* being the *occupancy number*. The average count per square is *m*, and *x!* is the product of all the numbers between *x* and 1 multiplied together. So that 5! is $5 \times 4 \times 3 \times 2 \times 1 = 120$. By convention 0! is taken as having the value 1, as does 1!

To take a simple example, consider the case of $m = 1$, which was done with

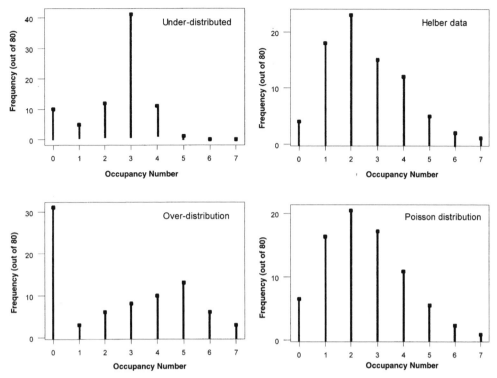

Figure 6.10 Frequency plots of 201 bacterial cells distributed over 80 grid squares, to illustrate under-distribution and over-distribution in comparison with the Poisson distribution and the actual experimental results

MINITAB in §6.4.1. With a pocket calculator, e^{-1} is 0.3679. Therefore the frequency of empty squares $(x = 0)$ is:

$$P_{x=0} = (0.3679).1^0/0! = 0.3679, \text{ or about } 37\%$$

Likewise the number of squares with three cells is:

$$P_{x=3} = (0.3679). 1^3/3! = 0.0613, \text{ or about } 6\%$$

These figures will be seen to agree with those produced by MINITAB in §6.4.1.

FN 6.2 Insights into chi-square

Along with the *t*-distribution and the *F*-distribution, the *chi-square* distribution is one of the most commonly employed for statistical tests of significance, its application being particularly with *count* and *proportion* data. The chi-square distribution itself involves a fairly complicated mathematical formula that will be found in advanced books on statistics and is beyond the scope of the present work. The output from the formula is a series of curves, one for each number of degrees of freedom. Each such curve is a plot of the probability density against the value of chi-square as produced by the equation already cited, viz:

$$\chi^2 = \Sigma (O - E)^2/E \qquad \text{Eq. 6.1}$$

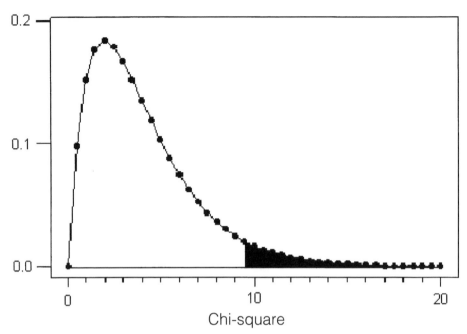

Figure 6.11   The chi-square distribution for 4 degrees of freedom. The shaded area encloses
the top 5% of the distribution and starts at chi-square = 9.49

For practical purposes, the information needed for an ordinary chi-square test is
the area in the extreme right-hand tail of the chi-square distribution curve, since
this corresponds to a particular probability. Figure 6.11 shows the chi-square
distribution for four degrees of freedom and plotted from MINITAB. It was made
by opening a new worksheet and in C1 with the *Make Patterned Data* command,
creating a set of numbers from 0 to 20 with 0.5 spacings. Then the corresponding
probability density values for chi-square were obtained by going to **Calc > Probability Distributions > ChiSquare**, and selecting the *Probability-Density* option, 4
Degrees of freedom and *Optional storage* in Col. 2. The shaded region is the 5% of
the area under the curve that starts at a chi-square value of 9.49 (see Table 6.9). The
use of MINITAB to obtain the conventional *P*-value from a found value of chi-
square has already been described (§6.6.2).

FN 6.3   Chi-square for exploring conformity to Poisson

This may be done by the method described for exploring differences between *Fields*
in the Helber chamber (Table 6.8). A new worksheet (Table 6.12) may conveniently be opened in the Helber project and the data on occupancy number for
the Poisson and the Helber distributions copied in to columns C1–C3 from Table
6.11. There is then the need for slight adjustment to fulfil the requirement for chi-
square that no *expected* number be less than about 5. For this reason the figures for
occupancy numbers 6 and 7 are added to those for occupancy number 5, to give the
values in C4 and C5. The chi-square calculation is then done as in §6.6.1 to deliver
the result in C6. This value is converted to its corresponding cumulative probability

Table 6.12 MINITAB worksheet for application of chi-square to determine whether the Helber slide data depart significantly from a Poissson distribution

C1 Occ No	C2 Pois	C3 Helb	C4 Pois Adj	C5 Helb Adj	C6 Chi Sq	C7 Cum Prob	C8 P(%)
0	6.49	4	6.49	4	1.92268	0.140263	86
1	16.3	18	16.3	18			
2	20.47	23	20.47	23			
3	17.14	15	17.14	15			
4	10.49	12	10.49	12			
5	5.41	5	8.81	8			
6	2.27	2					
7	1.13	1					

in C7 and finally into a conventional *P*-value, in percent, and with rounding to one decimal place, in C8. The high probability value of 86% means that the null hypothesis (that the two distributions do not differ by more than random-sampling fluctuations) is not challenged. Therefore the Helber data are confirmed as approximating closely to a Poisson distribution.

When the same procedure is applied to the under-distributed and over-distributed data sets, the differences from the Poisson are highly significant.

7

How to Deal with Proportion Data

Life is short, Art long, Opportunity fleeting, Experience treacherous, Judgement difficult.

Hippocrates, Aphorisms

7.1 DEVELOPING A SENSE OF PROPORTION

Proportion data arise from a wide variety of procedures, such as therapeutic trials of drugs and medical procedures, toxicity tests, seed germination experiments, water microbiology and genetical observations. The feature common to all of these is that the summarized results consist of *whole-number proportions*, such as 3/8 or 41/375. The raw data thus require each observational unit to be placed without ambiguity into discrete categories, such as dead/alive in a toxicity test, sterile/ contaminated in a sterility test, or male/female in data that record the distribution of the sex of individuals. Another name for such observations is *categorical data* because of each result being placed in a definite category. The common everyday example is the tossing of a coin, where the individual result is a head or a tail, and the outcome of a series of tosses is a proportion such as 7/10 for the number of heads in ten trials. Having obtained such data, some of the aspects to be explored by statistical analysis are:

- Significance of the differences of success rate in drug trials: is Treatment A better than Treatment B, based on comparing the success proportions?
- In sterility testing of injectable drugs, calculating the probability of a 'clear' test on a sample of, say, $N = 20$ ampoules, if the background incidence of contamination in the whole production batch is $X\%$.
- In analysis of water for microbial contamination, what is the most probable number (MPN) of organisms in tests where replicated culture tubes, inoculated with tenfold dilutions, stay clear or become turbid with microbial growth?
- In plant and animal genetical experiments, whether the ratios of phenotypic characters in the progeny correspond to Mendelian expectations?

With the count data of the last chapter, the Poisson distribution was taken as the underlying statistical model. Here, with proportion data, the binomial distribution provides the theoretical base. For comparing groups of data, the chi-square test will figure prominently with proportion data as it did with counts, and it is assumed that the reader will already have read about chi-square in the previous chapter. As with count data where the numbers are small, proportion data with small numbers require special treatment, and chi-square is not applicable in such circumstances (i.e. when an *expected* number is less than about 5).

This chapter does not consider antibody titres, where the result may be expressed as a proportion such as 1/128. Such data are best normalized by conversion to the logarithm of the reciprocal, e.g. 1/128 is converted to 128 and then into a $\log_{10}$ value of 2.107 (Chapter 5) or a $\log_2$ value of 7.0. Also dealt with elsewhere (Chapter 9) is the estimation of ED_{50} and LD_{50} values in dose–response titrations.

7.2 LARGE-GROUP PROPORTION DATA

7.2.1 Significance of differences

'Large' in the present context means groups of more than about 20, or at any rate of a size where neither the numerator nor the denominator of any of the proportions infringes the chi-square requirement for having no *expected* number less than about 5. As an example, the results of S. Banic (1975) are taken, in which injections of

Table 7.1 Data on the prevention of experimental rabies in guinea-pigs by vitamin C, set out as a 2 × 2 contingency table

Treatment	No. of deaths	No. of survivors	Total
Vitamin C	17	31	48
Control	35	15	50
Total	52	46	98

Reproduced by permission from The Editor of Nature, **258**, *153.*
Copyright © 1975 Macmillan.

vitamin C were investigated for their possible protective effect in guinea-pigs, all of which had been given the same challenge dose of rabies virus. There were 98 animals altogether, divided into two groups, and the final results of death and survival were as set out in Table 7.1. There were 17/48 deaths in those given vitamin C, compared with 35/50 deaths in the controls. Is this difference statistically significant?

Ignoring the marginal totals in the table, there are four central 'boxes' containing the numbers of dead and alive animals associated with the two treatments. This type of layout is known as a 2 × 2 *contingency table*, the 2 × 2 describing the relationship of the numbers (without totals) as being in two horizontal rows and two vertical columns.

This is a chi-square problem and to explore it with MINITAB, open a new MINITAB project and enter the data in the first three columns of a new worksheet as in Table 7.2. The first column has alphabetical information, which MINITAB recognizes by adding the suffix *T* for *Text* to C1. Alternatively the numbers 1 and 2 could have been used as codes for the two treatments, in which case MINITAB would have left the column as C1 (without the '-T'). The next two columns contain the figures for dead and alive. There is no need to add the row and column totals. Note particularly that the data are *not* entered in the form *result/total*, e.g. 17/48, but as the actual numbers of dead and alive. After entry of the data, save the Project as *Chap 7 rabies.mpj*.

To perform the chi-square test, go to **Stat > Tables > Chi-Square Test** and, when the dialog box (Figure 7.1) appears:

---

1) In *Columns containing the table:* insert the two data columns [Dead Alive] by highlighting both of them together in the left-hand list of columns and either double-clicking at the mouse or clicking on the *Select* button;

2) Click on [OK] and then go to the *Session* window to see the result (Table 7.3). If the *Project* is again saved at this point, all the material in the *Session* window will also be saved, along with the Worksheet.

MINITAB calculates chi-square and gives the final result in the last line of Table 7.3 as the *P*-value 0.001, or 0.1%. This allows firm rejection of the null hypothesis (that the difference between the two groups is due to sampling fluctuations), and

Table 7.2 MINITAB worksheet for rabies/vitamin C experiment

C1-T Treatment	C2 Dead	C3 Alive
Vit C	17	31
Control	35	15

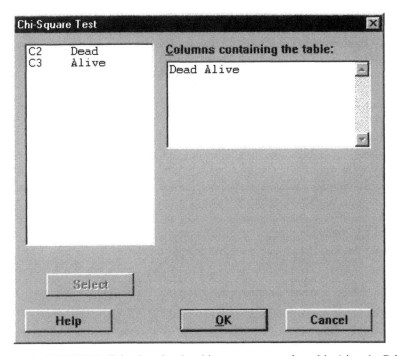

Figure 7.1 MINITAB dialog box for the chi-square test on the rabies/vitamin C data

Table 7.3 MINITAB *Session* window with the results of the chi-square test applied to the rabies/vitamin C data. Note that some cosmetic tidying up has been done to the printout, e.g. lines inserted in the table and the expected numbers put in italics

```
Chi-Square Test
Expected counts are printed below observed counts
```

	Dead	Alive	Total
1	17	31	48
	25.47	*22.53*	
2	35	15	50
	26.53	*23.47*	
Total	52	46	98

```
Chi-sq = 2.816 + 3.184 +2.704 + 3.056 = 11.760
d.f. = 1; P-Value = 0.001
```

therefore the difference in survival between the two groups of guinea-pigs is highly significant. The d.f. = 1, also on the last line, states that there is one degree of freedom. This is always the case with a 2×2 contingency table, irrespective of the size of the numbers in the table.

The main section of Table 7.3 is MINITAB's printout of the entered data, with row and column totals added. Entered in italics are the *expected numbers* under each *observed number*. These are calculated from the null hypothesis that only sampling fluctuations are responsible for the differences between the two groups. For example, suppose that vitamin C had no protective effect, then one can take the total 52 deaths out of the 98 animals infected and, say, that with a set of 48 (vitamin C group), the *expected* deaths would be $48 \times 52/98 = 25.47$. This is the *expected* number of dead in the vitamin C group, on the assumption of no protective (or enhancing!) effect. The same type of calculation is made for the other three compartments of the table. Then chi-square is calculated from Eq. 6.1, the first term being $(17 - 25.47)^2/25.47 = 2.816$, as in the second-last line of the MINITAB printout in Table 7.3. With d.f. = 1, the 5% point of chi-square is 3.84 (Table 6.9), the 1% point being 6.63 and the 0.1% point, 10.8. The output value of chi-square = 11.76 is greater than the 0.1% point and thus highly significant.

The formula for calculating chi-square in a 2×2 contingency table is given at the end of this chapter, in FN 7.1.

These results also illustrate a general point that a medical treatment (here vitamin C) gave experimentally a beneficial effect that statistically was 'highly significant', yet for practical purposes would not be particularly 'significant' or to be recommended. Thus from these results it would not be advisable for a person bitten by a rabies-infected dog to be treated only with vitamin C!

7.2.2 Confidence intervals of a proportion

In the rabies experiment, the dose of virus used for the challenge killed 35/50 of the control animals. This can also be expressed as 70%, or as the decimal fraction 0.7. It can also be regarded as an LD_{70} (*lethal dose in 70% of the test subjects*) quantity of virus. Wherever possible, it is good practice to attach confidence intervals to an experimental estimate, in order to have an idea of its range of uncertainty. Thus if further batches of guinea-pigs of theoretically the same sensitivity were given exactly the same dose of virus, would the result always be 35/50? Or might it range, say, from about 30/50 to about 40/50, due to random-sampling fluctuations? It is the same sort of problem as with an evenly balanced coin that is tossed 50 times. The result of successive 50-toss 'experiments' would not uniformly be 25 heads/50 tosses each time. The difference in the rabies experiment is that, unlike the coin, there is no information on the 'underlying' *true* death ratio with that particular dose of virus.

Use is made of the binomial distribution to obtain the probabilities of all the outcomes of challenging groups of 50 guinea-pigs, from the same stock and with the same dose of virus. The fact that 15 of the animals actually survived the challenge, means that in any future random set of 50 guinea-pigs (having the same range of sensitivity as those in the actual experiment) there is a theoretical possibility of having as few as *none* killed or as many as *all 50* killed. Therefore imagine having an infinitely large population of guinea-pigs of which 30% will resist challenge with that dose of rabies virus and 70% will be susceptible. MINITAB is then asked for the binomial probabilities of having 0, 1, 2, 3, ... 47, 48, 49 and 50 animals killed in a random set of 50. This can be done on the same worksheet as the chi-square test since there is plenty of space available from Col. 4 on the right.

The first step is to label C4 with *Dead/50* and put the numbers from 0 to 50 in it. This is done by going to **Calc > Make Patterned Data > Simple Set of Nos** and, in the dialog box, entering:

---

1) *Store patterned data in:* 'Dead/50';
2) *From first value:* $\boxed{0}$
3) *To last value:* $\boxed{50}$
4) *In steps of:* $\boxed{1}$
5) *List each value:* $\boxed{1}$ *times*
6) *List the whole sequence:* $\boxed{1}$ *times*
7) $\boxed{OK}$.

To find the binomial probabilities for each of these 50 death rates, label the next free column C5, with *Bin P(0.7)*, as shorthand for *binomial probability for a distribution with population frequency 0.7*. Then go to **Calc > Probability Distributions > Binomial** and, when the dialog box opens, make the entries as:

---

1) Select • *Cumulative probability* from the three choices offered;
2) *Number of trials:* $\boxed{50}$;
3) *Probability of success:* $\boxed{0.7}$, which is the proportion of deaths in the control group;
4) *Input column:* $\boxed{\text{'Dead/50'}}$;
5) *Optional storage:* $\boxed{\text{'Bin P(0.7)'}}$;
6) $\boxed{OK}$.

This will yield the two columns of figures shown (in a space-saving format with four columns) in Table 7.4. The *lower 95% confidence interval* is the point in the distribution where the cumulated probability reaches 2.5% or 0.025. This occurs with a Dead/50 value of 28 and is highlighted. Likewise the upper 95% confidence interval is where the cumulated probability reaches 97.5% or 0.975. This is in-between the Dead/50 values of 40 and 41 where the cumulated probabilities are 0.96 and 0.98, but closer to the latter, which would therefore be described as an *approximate upper 95% CI*. These intervals can now be attached to the value of the LD_{70}. The assumed population probability of 0.7 was based on a Dead/50 number

Table 7.4 Cumulative binomial probabilities for different numbers of deaths out of 50, where the assumed population death frequency is 0.7. The boxed values in **bold** mark the approximate lower and upper 95% confidence intervals of the frequency estimate. To save space, the first 19 entries, which were all 0.0000 in C5, and the last three, which were all 1.0000, have been omitted

C4 Dead/50	C5 Bin P(0.7)	C4 Dead/50	C5 Bin P(0.7)
19	0.00000	34	0.43082
20	0.00001	35	0.55317
21	0.00004	36	0.67212
22	0.00012	37	0.77713
23	0.00034	38	0.86096
24	0.00093	39	0.92115
25	0.00237	40	0.95977
26	0.00559	**41**	**0.98175**
27	0.01228	42	0.99274
28	**0.02509**	43	0.99751
29	0.04776	44	0.99928
30	0.08480	45	0.99983
31	0.14056	46	0.99997
32	0.21781	47	1.00000
33	0.31612		
continued top of right-hand column			

= 35, which was also described as an LD_{70} dose. Thus the 95% CIs of this LD_{70} are 2×28 and 2×41, or 56 and 82. Therefore the challenge injection of rabies virus given to the guinea-pigs can be described as an LD_{70} dose, with 95% CIs of 56 and 82.

7.2.3 Effect of group size

For reasons of financial cost, and frequently also of ethical considerations, it is necessary to choose carefully the group size to explore a particular experimental effect in humans or animals. For example, the effect of vitamin C in the rabies experiment was highly significant, since it gave $P = 0.001$, or 0.1%. However, a total of 98 animals was used and the question arises of whether, with hindsight, it would have been feasible to have had smaller groups and accepted a smaller chi-square and a higher (i.e. *less significant*) P-value.

This can be explored as an exercise on the worksheet by progressively reducing the size of the numbers in the 2×2 table and doing a new chi-square test after each change. MINITAB will not accept decimals in a 2×2 table; therefore the nearest whole numbers should be used, while keeping the survival ratios in the two groups as close as possible to 35% and 70%. This has been done in Table 7.5, which starts with the original groups and progressively reduces them to approximately one-half, one-third and one-fifth of their initial sizes. The half-size experiment with groups of

Table 7.5 Effect of group size on the outcome of the chi-square test in the rabies experiment

Data in relation to original	Group	No. of animals Dead	Alive	Total	Death rate (%)	Chi-square	P
Unaltered	Vit C	17	31	48	35.4		
	Control	35	15	50	70.0	11.8	0.001
Approx half-size	Vit C	8	15	23	34.8		
	Control	17	7	24	70.8	6.13	0.013
Approx one-third	Vit C	6	10	16	37.5		
	Control	12	5	17	70.6	3.64	0.056
Approx one-fifth	Vit C	3	6	9	33.3		
	Control	7	3	10	70.0	2.55 *	0.11

* In delivering this value of chi-square, MINITAB reported that three cells have expected numbers less than 5.

23 and 24 animals gave a *P*-value of 0.013, which is *significant*, but not *highly significant*. However, reduction to one-third size, with groups of 16 and 17 animals, yielded a *P*-value that was not quite significant (by the conventional criterion of having to be 0.05 or less). Moreover with such group sizes, the addition of one animal to a cell can make the result significant. For example, adding an extra death to the control group, to make it 13/18 (72.2%) instead of 12/17 (70.6%), increases the chi-square to 4.142, with $P = 0.042$ (now *significant*).

Further reduction of group sizes to 9 and 10, about one-fifth of original, gave a *P*-value of 0.11, which is not significant. These small groups contained *expected* numbers less than 5, which MINITAB duly reported as a cautionary note, meaning that the use of chi-square for such small numbers is unsatisfactory.

The above calculations indicate how, with hindsight, the group size could have been reduced somewhat without obscuring the protective effect of vitamin C, but this kind of information is not so useful for forward planning. For example, in a repeat of the original experiment, sampling fluctuations alone might reduce the *apparent* lethality of the virus challenge in the control group, while coincidentally *decreasing* the apparent protective effect of the vitamin C. Thus if the challenge dose was operating at its lower 95% confidence interval, then only 28, instead of 35, control animals would have died. Likewise suppose that, instead of 17 dying in the vitamin C group, 19 died. Then chi-square with the same group totals is 2.64, giving a *P*-value that is no longer significant. Thus the vitamin C protective effect disappears and would not have been detected.

Such considerations lead to a much more sophisticated approach to choosing group sizes. Account can be taken of the magnitude of the effect being sought, the background incidence in the control population, and what probability one wants to specify for not missing a result significant at a particular *P*-value. This whole area is known as *power and sample size* statistics and is well developed on MINITAB Release 12. Its application to the rabies data is worked through in FN 7.3.

7.2.4 Relative risks and their confidence intervals

It is common for clinical trials to yield proportion data with the same layout of four compartments, as in the rabies/vitamin C experiment. In addition to analysis of significance of differences, it may be useful to calculate *relative risks* together with confidence intervals. Thus in the rabies experiment, the vitamin C reduced the dead/total ratio from 35/50 to 17/48, or from 70% down to 35.4%. The figures can alternatively be expressed as a relative risk. Here the relative risk is that for an otherwise untreated guinea-pig to contract rabies from the dose given, as compared with one which had received vitamin C. To generalize the data, it is convenient to express the 2×2 contingency table (Table 7.6) with letters to represent the numbers in each compartment.

Table 7.6 General format of a 2×2 contingency table

Group feature	No. of individuals		Total
	Responding	Not responding	
A (Present)	a	b	$(a + b)$
B (Absent)	c	d	$(c + d)$
Total	$(a + c)$	$(b + d)$	N

The relative risk, or risk ratio, R is then given by:

$$R = \frac{c/(c + d)}{a/(a + b)}$$ Eq. 7.1

$$R = \frac{35/50}{17/48} = 0.7/0.354 = 1.98 = \sim 2.0$$

In other words. without the vitamin C, there was an approximately doubled chance of dying from that particular challenge dose of rabies virus.

To find the 95% confidence intervals of R requires a pocket calculator for the use of natural logarithms (Ln) and is calculated in several steps, as described by Morris and Gardner (1998). Thus the value of $R = 1.98$ is converted into its natural log as Ln(1.98) = 0.6831.

The *standard error of the natural log of R*, abbreviated as *SE(LnR)*, is given by:

$$SE(LnR) = \sqrt{[1/a] - [1/(a + b)] + [1/c] - [1/(c + d)]}$$ Eq. 7.2
$$= \sqrt{0.0588 - 0.0208 + 0.0286 - 0.0200}$$
$$= \sqrt{0.0466} = 0.2159$$

The Ln values of the lower (L) and upper (U) 95% CIs are then given by:

$$Ln\ (95\%\ CI) = LnR \pm 1.96\ SE(LnR)$$
$$= 0.6831 \pm 1.96 \times 0.2159 = 0.6831 \pm 0.4232$$
$$= 0.2599\ and\ 1.1063$$

Taking antilogs, with the e^x button on the calculator gives the 95% confidence intervals as 1.3 and 3.02. Thus the effect of *not giving* the vitamin C can be summarized as involving a relative risk of 2.0, with 95% CIs of 1.3 and 3.0. In interpreting such results it should also be borne in mind that if the lower 95% limit had not been above 1.0, the apparent protective effect of the vitamin C would not have been significant at the $P = 5\%$ level.

By reversing the presentation of the groups in the table, the effect of the vitamin C could alternatively have been expressed as *reducing* the risk of rabies to a ratio of 0.5 and with correspondingly altered confidence intervals.

7.3 SMALL-GROUP PROPORTION DATA

As emphasized above, chi-square ceases to be applicable when group sizes become small, the critical requirement being that no *expected number* be less than about 5. An alternative approach is therefore required for exploring the significance of differences between small groups. This can be done at various levels of mathematical input, but the most direct is to consult pre-calculated statistical diagrams. A selection of these is provided in Appendix A3 at the end of the book. It allows determination of significance of differences in a series of group sizes from three up to 15 but assumes that the two groups being compared are of equal size. As an example, a 10×10 comparison is presented in Figure 7.2. Suppose the data sets are 4/10 and 8/10, are these significantly different? The procedure is to enter the numerator of one of them, e.g. the 4, on the top line and read down vertically to the point of intersection with the 8 entered from the vertical axis. In this case the intersection is on a clear square, which is in the zone of P-values greater than 5%, therefore the difference in the two ratios is not significant. On the other hand, 4/10 and 9/10 would be significantly different at the $P = 5\%$ level and 4/10 and 10/10 at the $P = 1\%$ level.

Many years ago, Finney *et al.* (1963) produced a set of *Tables for Testing Significance in a 2 × 2 Table*. This allows comparisons of groups of unequal size up to 40 in the denominator.

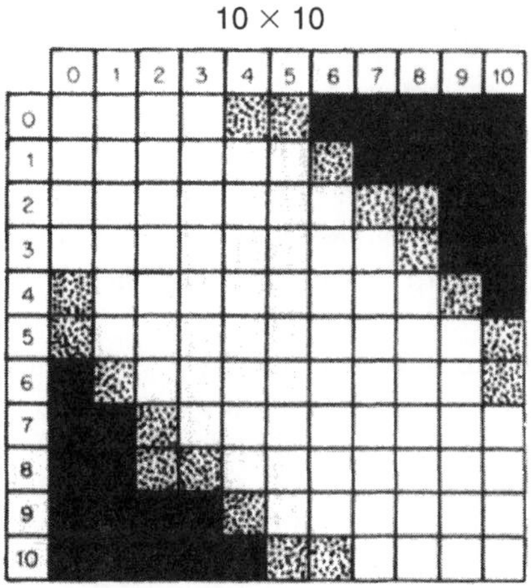

10 × 10

Figure 7.2 Probability diagram for determining the significance of differences between two proportions, each with denominator 10. Black = significant at the $P \leq 1\%$ level; dotted = significant at the $P = \leq 5\%$ level, and clear = not significant $(P > 5\%)$

7.4 ARE THOSE AMPOULES STERILE?

In the pharmaceutical industry, sterility tests are performed to ensure that products labelled 'sterile' are free from living microbes. Once a batch of an injectable substance has been dispensed into its final containers, the law requires that 2%, or 20, of the containers, whichever number is smaller, be selected at random and tested for sterility by inoculation into culture media, which are then incubated and inspected for development of turbidity. The question here is to calculate the probability that a 20-sample test might be passed as 'satisfactory if, in fact, $X\%$ of the containers were contaminated, but were not among those chosen for testing'.

This is a problem involving the binomial distribution. First prepare the continuation of the previous worksheet by labelling columns C8–C14 as in Table 7.7. C8, labelled a, contains the integers 0 to 5 for the possible numbers of contaminated containers that might be present in a random sample of 20 containers from the production batch. C9–C11 are labelled to receive the binomial probabilities for each value of a, assuming three levels of contamination in the batch, viz, 0.1 (10%), 0.01 (1%) and 0.001 (0.1%). C12–C14 contain the values delivered by MINITAB into C9–C11, after conversion into percentages and rounded to one decimal place. To do the calculations, go to **Calc > Probability Distributions > Binomial** and, when the dialog box appears opens, proceed as follows:

---

1) Select • *Probability* from the three choices offered;
2) *Number of trials:* $\boxed{20}$;
3) *Probability of success:* $\boxed{0.1}$, to examine the assumption that 0.1, or 10%, of the production batch is contaminated;
4) *Input column:* $\boxed{a}$;
5) *Optional storage:* $\boxed{\text{'X = 0.1'}}$;
6) $\boxed{OK}$.

Repeat this process with 0.01 and 0.001 successively at Step 3 and with delivery to *Optional storage* in $X = 0.01$ and $X = 0.001$ respectively. Then to make the binomial probabilities easier for the eye, convert them to percentages and rounded to one decimal place by opening *Calculator* and using the function *ROUND(100*'X = 0.1', 1)* for C9, and similarly with the other column headings *for $X = 0.01$ and $X = 0.001$*.

The important parts of the output in Table 7.7 are highlighted in bold. Starting from the right, it shows (C14, line 1) that there is a 98% probability of failing to detect contamination ($a = 0$) when a 20-container test is done on a production batch with an 0.1% contamination rate, i.e. an average of 1 contaminated container per 1000. This means that in a batch of 100000 containers, there could be 100 of them contaminated but the chance of detecting this contamination is very small. When the contamination rate is 1% (C13) there is an 82% chance of failing to detect contamination and only a 16.5% chance of detecting one contaminated container. However, when the contamination rate is 10% (C12), there is only a 12%

Table 7.7 MINITAB worksheet for calculating binomial probabilities for sterility testing and expressing them in percentages. The results in **bold** are the percentage probabilities of *failing* to detect contamination in a 20 sample test if the background contamination rates are $X = 0.1, 0.01$ and 0.001, i.e. 10%, 1% and 0.1% contamination

C8	C9	C10	C11	C12	C13	C14
a	$X = 0.1$	$X = 0.01$	$X = 0.001$	$X = 0.1$ (%)	$X = 0.01$ (%)	$X = 0.001$ (%)
0	0.121577	0.817907	0.980189	**12.2**	**81.8**	**98**
1	0.270170	0.165234	0.019623	27.0	16.5	2
2	0.285180	0.015856	0.000187	28.5	1.6	0
3	0.190120	0.000961	0.000001	19.0	0.1	0
4	0.089779	0.000041	0.000000	9.0	0.0	0
5	0.031921	0.000001	0.000000	3.2	0.0	0

probability of *failing* to detect the contamination, leaving about an 88% probability of detecting one or more contaminated containers.

These calculations are applicable to other manufacturing processes where there is the possibility of defective units arising and where a random-sampling procedure is in place to control the quality of the output.

7.5 MOST PROBABLE NUMBERS IN MICROBIAL COUNTING

There are numerous methods for determining the numbers of viable microbial cells in a sample, the classic procedure being to spread a known volume of the sample on the surface of a culture plate, incubate and count the numbers of colonies that appear. This depends on the assumption that each viable cell gives rise to one colony. The statistical analysis of such data is discussed in the previous chapter on *Count data*. Another procedure that may be used when colony counting is not applicable, is the *most probable number (MPN)* method. The principle of this method is to make serial dilutions, usually tenfold, of the sample and inoculate constant volumes into replicated tubes of sterile, liquid growth-medium that are then incubated and inspected for microbial growth. Any tube receiving one or more viable microbial cells will exhibit growth, while the rest remain sterile. The results of such a test are then a set of *proportion data*, such as 4/5, 2/5 and 0/5 for the proportions of positive tubes inoculated respectively with the sample undiluted, and at 1/10 and 1/100 dilutions. These data do not yield an exact count, but instead *an estimate of the most probable number* of viable cells per unit volume that was inoculated from the undiluted sample. A further point is that, normally, a wider range than three dilutions would be set up, to make sure of reaching an end-point, but only three of the dilutions would be likely to yield *informative triplets*. These latter are the results from three successive dilutions that are not either all positive for growth, or all negative.

The MPN value is most readily obtained from a statistical table such as that in Appendix A1 (due to Taylor, 1962), rather than by individual calculation for each set of results. For the example above, the Table A1 gives the MPN as 2.2 viable cells per inoculation volume used in the undiluted sample. An additional table, A2, is provided for interpreting the results of similar sets of eight replicate tubes at each of three tenfold dilutions (Norman and Kempe, 1960). A procedure involving twofold dilutions of the sample in microtitre plates was described by Rowe *et al.* (1977), who also provided tables of MPN values and SEMs.

MPN procedures have the drawback of inherently low precision, as expressed by the wide confidence intervals. Cochran (1950) produced a table of *confidence-interval (CI) factors* for MPN tests with various dilution series and various numbers of inoculated tubes at each dilution. For a five-tube test with tenfold dilutions and three informative triplets, the factor is 3.30 and for a similar eight-tube test it is 2.57. The limits are obtained by dividing and by multiplying the MPN by the CI factor. Thus for the example under consideration, the lower CI is 2.2/3.30 and upper CI is 2.2 × 3.3. These give CIs of the MPN as 0.67 and 7.26, an 11-fold range that illustrates the imprecision of the MPN method.

The fact that MPN methods are used at all is because they may be the only ones possible with some microorganisms in certain types of sample. For example, portions of muddy river water being examined for coliform bacteria may not be suitable for plating on to solid culture media to obtain a colony count, and an MPN method may be an acceptable alternative.

7.6   IS THAT INHERITANCE MENDELIAN?

Another example of proportion data is provided by the results of breeding experiments where the offspring are recorded unambiguously as belonging to discrete categories, such as male/female. For instance, in the human ABO blood group system there are three major alleles, A, B and O, each of which can occupy the ABO locus on each haploid chromosome. In each human diploid cell there are two homologous loci, so that the possible genotypes are AA, AO, AB, BB, BO and OO. Since the O gene is 'silent' and the A and B genes are co-dominant, the corresponding blood group phenotypes are A, AB, B and O. In former years there was much controversy about this scheme and, in one investigation, the blood groups of the offspring of marriages in which both parents were AB was studied. Here (Weiner, 1943) the *observed* numbers of children with particular blood groups were:

Group A	28
Group B	36
Group AB	65
Total	129

According to the theory discussed, the predicted distribution of blood groups in offspring from AB × AB matings is:

Group A	1/4
Group B	1/4
Group AB	1/2

Therefore in a population of 129 individuals, the *expected* numbers would be:

Group A	32.25
Group B	32.25
Group AB	64.5

It can be seen without doing any statistical tests that the *observed* and *expected* numbers are in close agreement. Nevertheless to confirm this impression, a chi-square test will be performed with Eq. 6.1 and using Table 6.9 to interpret the result. The three *observed* numbers of Group A = 28, Group B = 36 and Group AB = 65 are subtracted from their *expected* numbers of respectively 32.5, 32.5 and 64.5, the differences squared and divided by the expected numbers, to give chi-square from Eq. 6.1 as:

$$\chi^2 = (28 - 32.25)^2/32.25 + (36 - 32.25)^2/32.25 + (65 - 64.5)^2/64.5$$
$$= 1.00$$

As with all chi-square tests there is a null hypothesis which states that the apparent difference between observed and expected values is due to sampling fluctuations. This hypothesis is threatened with being overturned only when the found chi-square exceeds the tabulated value at the $P = 5\%$ level (chi-square for 2 d.f. = 5.99). Therefore the differences between observed and expected numbers in the present data are well within sampling fluctuations and the null hypothesis stands. Thus the inheritance pattern expected with two co-dominant genes was fulfilled.

In his original work on the inheritance of structural characteristics in garden peas Mendel observed that when tall peas were intercrossed, approximately one-quarter of the progeny were short and approximately three-quarters were tall. This and similar observations gave rise to the theory of inheritance of dominant and recessive characters. The chi-square test did not exist in Mendel's time but if it had it could have been used to determine the closeness of his observations to the theoretical expectations. The one-quarter incidence of short peas in the progeny of intercrossing tall peas suggests that the latter can be represented genetically as *Tt*, where *T* is the dominant gene that confers tallness and *t* is the recessive allele which is only manifested if present at both loci. Therefore the crossing of *Tt* × *Tt* gives *TT*, *Tt*, *tT* and *tt*, of which only the last is short and comprises one-quarter of the progeny. Mendel's actual results were 787 tall and 277 short peas in a total of 1064 progeny. The corresponding *expected* numbers are 0.75 × 1064 = 798, and 0.25 × 1064 = 266. Application of chi-square as before gives:

$$\chi^2 = (787 - 798)^2/798 + (277 - 266)^2/266$$
$$= 0.61$$

This has one degree of freedom and, since it is much less than the tabulated chi-square ($\chi^2 = 3.84$ for $P = 5\%$) for 1 d.f., supports the theory of three-quarters and one-quarter for the frequencies of expression of dominant and recessive characters in a heterozygous cross.

7.7 SUMMARY

This chapter differs from earlier ones in taking data from diverse sources to illustrate the statistics that may be applied to the subject under discussion – *proportion* data. A drug trial was used as an example of a 2 × 2 contingency table to explore the effect of a treatment on a disease. This led to calculating the confidence intervals of proportions and also (below) into *Power* calculations to determine the necessary sizes of experimental groups for the effects expected. Other examples of proportion data were taken from sterility testing of injectable drugs, microbial counting by the *most probable number (MPN)* method, and the examination of outcomes of genetical experiments. Underpinning these examples were the binomial distribution and the chi-square test. For proportions of numbers that are too small for chi-square analysis, special diagrams were provided to provide graphical solutions. The same approach was used for MPN data.

FURTHER NOTES

FN 7.1 Chi-square for 2 x 2 contingency table

The 2 × 2 contingency table was described in §7.2 and Tables 7.1 and 7.6 but, since MINITAB performed the chi-square calculation automatically, there was no need to use any equations. For reference, however, there are two equations for calculating chi-square for a 2 × 2 contingency table, one being 'better', i.e. more exact, than the other. With the generalized 2 × 2 contingency table as defined in Table 7.6, the 'better' equation is:

$$\chi^2 = \frac{N\{|\,ad - bc\,| - \tfrac{1}{2}N\}^2}{(a + b)\,(c + d)\,(a + c)\,(b + d)} \qquad \text{Eq. 7.3}$$

The straight vertical brackets on the top line indicate that the products ad and bc are to be subtracted so as to yield a positive answer. Substituting the rabies data from Table 7.1 gives:

$$\chi^2 = \frac{98\{|\,17 \times 15 - 31 \times 35\,| - \tfrac{1}{2}.98\}^2}{48 \times 50 \times 52 \times 46}$$
$$= 10.41$$

This is not exactly the same as the chi-square = 11.760 given by MINITAB at the bottom of Table 7.3 with the same input data. The explanation, as stated in the MINITAB *User's Guide*, is that *Yates's correction* is omitted from the procedure used by MINITAB. This factor is the $\tfrac{1}{2}N$ on the top line of Eq. 7.3. When this factor is left out, the expression becomes Eq. 7.4:

$$\chi^2 = \frac{N(ad - bc)^2}{(a + b)\,(c + d)\,(a + c)\,(b + d)} \qquad \text{Eq. 7.4}$$

Substitution of the data from Table 7.1 gives:

$$\chi^2 = \frac{98(17 \times 15 - 31 \times 35)^2}{48 \times 50 \times 52 \times 46}$$
$$= 11.76$$

This is the same result as provided by MINITAB. The difference between the two formulae is not great and would be of little importance in most instances. However, purists would say that the use of Yates's correction gives the better result.

FN 7.2 Formula for the binomial distribution

In §7.4, MINITAB was used to supply the binomial probabilities (P) for finding the various numbers (a) of contaminated vials in random sets of 20 (N) and assuming various levels of contamination (x) in the production batch. The same information could otherwise be obtained from the binomial formula:

$$P = \frac{N! \; (X)^a \cdot (1-X)^{N-a}}{a! \; (N-a)!}$$

Eq 7.5

To see how this works, let us take the case of a production batch with a contamination rate (X) of 0.01 (i.e. 1%) and a sample size of $N = 20$ containers taken at random for testing. The probability ($P_{a=0}$) of *failing to detect* (therefore put $a = 0$) contaminated ampoules in the sample taken is:

$$P_{a=0} = \frac{20! \cdot (0.01)^0 \cdot (1-0.01)^{20}}{0! \; (20-0)!}$$

The 20! means $20 \times 19 \times 18 \times 17 \times \ldots \times 3 \times 2 \times 1$ but does not have to be worked out since it cancels with the $(20-0)!$ on the bottom line. The $(0.01)^0$ equals zero since any number raised to the power zero is equal to 1. Therefore the calculation simplifies to:

$$P_{a=0} = (0.99)^{20} \text{ which on a calculator with a } y^x \text{ button yields } 0.818,$$

as in C10, top line, of Table 7.7.

Likewise, the probability of detecting *one* contaminated ampoule is obtained by inserting $x = 1$ and cancelling the top and bottom factorials to give:

$$P_{a=0} = \frac{20! \cdot (0.01)^1 \cdot (1-0.01)^{19}}{1! \; (20-1)!}$$

$$= \frac{20 \times 0.01 \times 0.99^{19}}{1}$$

$$= 0.165, \text{ the same as the } 0.165234 \text{ in C10 of Table 7.7.}$$

FN 7.3 Power calculations for group size

In §7.2.3 and Table 7.5, the effect of group size on the outcome of the chi-square test on the rabies data was explored. As the group size was reduced from about 50, while keeping the proportion of deaths the same, the chi-square value for the protective effect of vitamin C progressively diminished and, with groups of about 16 or less, became non-significant.

MINITAB allows a more sophisticated approach to calculating the appropriate size of experimental groups by *power* calculations. These can be done both for proportion data as in the rabies experiment, and with measurement data, as in the pipetting experiment of earlier chapters. To follow the procedure, it is necessary to have a good understanding of the null hypothesis in its application to two treatment groups, as in the rabies experiment. Some mental gymnastics are required in what follows because of the double negatives. (Note the omission in what follows of the words much loved by writers of statistics and computer texts: *easy, simple, straightforward, clearly.*) But do not be discouraged!

Before applying the power calculations to the rabies data, consider first the general hypothetical case of a *treatment* proportion and a *control* proportion where the chi-square test yielded the critical 5% point of 3.84. This is the tabulated chi-square for 1 d.f. and $P = 5\%$, and would result in rejection (just!) of the null hypothesis. The difference between the two groups would be declared 'statistically significant at the $P = 5\%$ level'. But in coming to this conclusion, it has to be remembered that a chi-square of 3.84 *could* arise on 5% of occasions *purely because of* random sampling fluctuations, and *not* because the two groups are actually different. That is, if there was *one* underlying population from which repeated random samples were taken, so as to get repeated sets consisting of one 'treatment' group and one 'control' group, then on 5% of occasions in a long run series, a chi-square of 3.84, or greater, should emerge. This would result in a *false* declaration that the two groups were significantly different when in fact they weren't. This sort of mistaken conclusion is called an *error of the first kind*, or *type 1 error*.

A *type 2 error* is where there are two groups that *actually* came from *different* underlying populations and where there was *failure* to detect the difference, that is, the chi-square was less than 3.84. This could occur if, by a fluke of random sampling, data were gathered from the *lower end* (delivered by sampling fluctuations) of the *truly higher* population and compared with the *higher end* of the *truly lower* population.

The purpose of the power calculations is to take note of these possibilities, and particularly to avoid *type 2 errors* (i.e. failing to find a difference that really exists and which matters to the experimenter). Thus a *type 2 error* could result in a useful drug being wrongly discarded because of apparently not showing a significant beneficial effect (due to a fluke of random sampling).

To do the power calculation with MINITAB it is necessary to specify:

- The desired numerical size of the effect being investigated, like wanting to reduce by 50% the death rate in rabies-challenged guinea-pigs by giving a possibly protective substance;
- (The hardest to understand!): The probability of *avoiding a type 2 error*. That is, the probability of failing to discover an effect that is really there, and which matters to the experimenter.
- The *P*-value that is to be applied in the null hypothesis (usually $P \leq 5\%$).
- Whether, in comparing the treatment group (T) with the control (C), there is interest in:
  a) C being significantly *less than* T;
  b) C being significantly *different from* T;
  c) C being significantly *greater than* T.

MINITAB uses the symbol α for the probability of making a type 1 error (when the null hypothesis is true, but through a sampling fluctuation is rejected); α is also described as the *level of significance*. Correspondingly, β is the probability of making a type 2 error (when the null hypothesis is false, yet is not rejected).

The technical term *power* is defined as $(1 - \beta)$, and is the *probability of correctly rejecting the null hypothesis when it is false*. It is desirable to have a high power to detect a difference that is important, and a low power for differences that are inconsequential.

Now for MINITAB, to see how it can be applied to the rabies data: Go to **Stat > Power and Sample Size > 2 Proportions** and, in the dialog box, enter the values as shown in Figure 7.3. The top section is the main box and the lower section is the

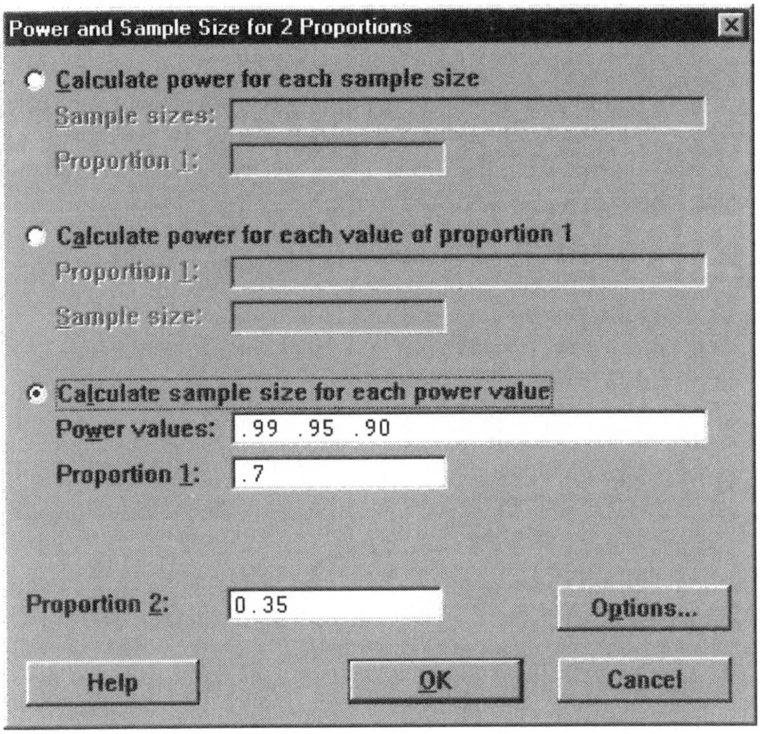

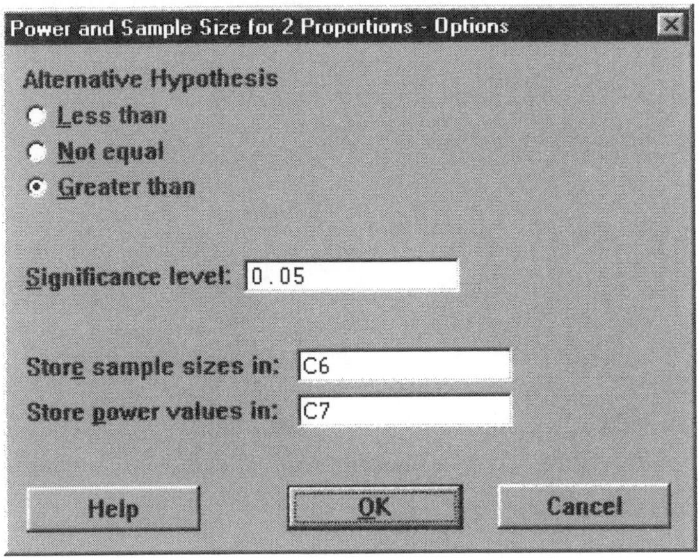

Figure 7.3 MINITAB dialog box for a *Power* calculation on the rabies/vitamin C data. Top, main box; bottom, *Options* sub-dialog box

internal box that is opened by selecting *Options* on the first. It will be noted that the proportions 0.7 and 0.35 have been inserted as proportions 1 and 2, in conformity with the death rates in the rabies experiment, and power values of 0.99, 0.95 and 0.90 chosen as reasonable requests. In the internal window, the (null) hypothesis was chosen as (c) above, and the default entry of 0.05 for α was left unaltered. Columns 6 and 7 on the worksheet were identified as destination columns for the output, which MINITAB also delivered to the Session window as:

```
Power and Sample Size
Test for Two Proportions
Testing proportion 1 = proportion 2 (versus >)
Calculating power for proportion 1 = 0.7 and proportion 2 = 0.35
Alpha = 0.05; Difference = 0.35
Sample    Target    Actual
  Size    Power     Power
    58    0.9900    0.9907
    40    0.9500    0.9525
    32    0.9000    0.9059
```

This showed that group sizes of 58, 40 and 32 are required for powers of 0.99, 0.95 and 0.90 respectively, and for the other features assumed in the calculation. It therefore appears that the group sizes of 48 and 50 actually used in the experiment were of the right order for confident ($P = 95$–99%) detection of the effect reported. For further familiarization with the power calculations, it is worthwhile to try out other combinations of assumed features in the data. The Target Power and Actual Power in the output are given because the sample sizes have to be whole numbers.

8

Correlation and Regression

It is vain to do with more what can be done with fewer.

William of Occam (early 14th century)

8.1 PRELIMINARY SIGNPOSTING

8.1.1 Data set for correlation and regression

Are students consistent in their exam results from year to year? More specifically, to what extent are university students' marks and ranking in the class, at the end of the fourth undergraduate year, related to the same students' results at the end of third year? Some sample data, are in Table 8.1, where the total percentage marks of a group of 22 students at the end of their third and fourth years are given. The students have been given identification (ID) numbers in order of merit from the top, based on the fourth-year results. The layout is transcribed from the first three columns of a MINITAB Worksheet within a MINITAB Project that has been saved as *Chap 8 MINITAB.mpj*.

From a statistical standpoint, each *observational unit* (student) is described by two independent variables – a third-year measurement and a fourth-year measurement. Such data provide convenient material for both *correlation analysis* and *regression analysis*. Neither procedure, of course, functions as a crystal ball for foretelling the future! Thus neither allows an exact prediction about how the individual student at the end of third year will necessarily perform at the end of fourth year, but overall trends for the class can be established and probability statements can be made.

Table 8.1 Transcript of MINITAB worksheet with examination results of a group of 22 students at the end of their third and fourth years at university. The listing is in order of merit of the fourth-year results

C1 Student	C2 3rd Year	C3 4th Year
1	76.8	72.9
2	72.7	70.7
3	75.6	70.0
4	63.9	68.5
5	68.2	67.4
6	62.6	66.2
7	61.2	65.7
8	65.4	64.9
9	60.1	64.2
10	57.6	63.7
11	53.7	63.4
12	65.8	63.1
13	65.2	62.8
14	62.2	62.4
15	61.4	61.3
16	53.1	60.3
17	67.2	59.9
18	50.4	59.6
19	56.9	57.8
20	53.7	55.4
21	51.8	53.8
22	58.7	50.9

8.1.2 Scope of correlation

Correlation analysis determines only the extent to which the two variables attached to each observational unit are *associated*, i.e. go hand-in-hand together. The output is a *correlation coefficient*, which is a number within the range from minus 1.0 through zero to plus 1.0. The coefficient, to be of full use, needs a *P*-value for the *significance* of the correlation. A point to note is that there are several different correlation coefficients, identified by the originator's name (Pearson, Spearman and Kendall). Also there are *partial* and *multiple* correlations where additional variables can be included.

8.1.3 Scope of regression

Regression analysis starts with different assumptions. First, one of the sets of measurements has to be identified as the *independent* variable (x) and the other as the *dependent* variable (y). In *linear* regression, the equation for the best-fitting *straight* line is calculated and fitted on the plotted points. There is also the opportunity for much additional numerical output in regression analysis as detailed below.

8.1.4 Graphical plot

Before starting either correlation or regression analysis, it is useful to do a *point plot* (§1.7.1) of the data, to have a general visual impression of the material under investigation. This is shown in Figure 8.1 as a scatter diagram, with each student represented by one point. The plot of fourth-year mark, on the *y*-axis, against third-year mark on the *x*-axis shows a general upward trend from left to right, indicating that there was a tendency for students' fourth-year results to be related to the performance in the third year (i.e. a positive correlation), but superimposed on the

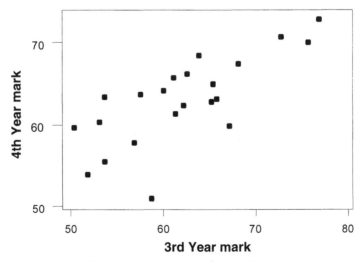

Figure 8.1 Point plots of third- and fourth-year marks of 22 university students, as an example of correlation

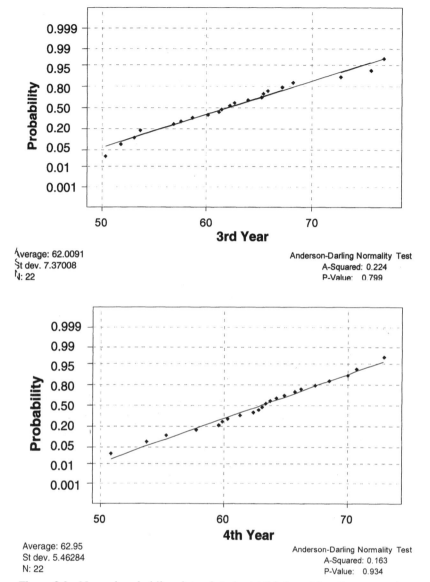

Figure 8.2 Normal probabilty plots of students' third- and fourth-year marks

general trend there is much scatter. Correlation analysis investigates the extent of the directional trend in relation to the scatter.

8.2 CORRELATION OF NORMALLY-DISTRIBUTED VARIABLES

8.2.1 Test for normality

The first step in correlation analysis is to decide whether or not the two variables are normally distributed. If there is reason to believe that *both* of them are, then the *product–moment correlation coefficient* (also known as Pearson's correlation coefficient *r*) is calculated. Using the method already described for normality testing (§3.4), Figure 8.2 shows that both the third-year and the fourth-year marks fulfil the criterion. However, if either or both variables had been non-normal, then one of the 'rank correlation coefficients' (Kendall or Spearman) would be appropriate. The latter are examples of non-parametric statistical procedures, i.e. where no assumption is made about the shape of the distribution of the underlying populations from which the data were drawn. Such analyses are based on *ranking*, and the Kendall procedure is described in §8.3.

8.2.2 Correlation on MINITAB

To carry out correlation analysis with normally-distributed variables on MINITAB, go to **Stat** > **Basic Statistics** > **Correlation** and, when the dialog box appears, enter the commands as in Figure 8.3:

1) *Variables:* '4th Yr' '3rd Yr';
2) ☑ *Display p-values*
3) OK.

The MINITAB output will be found in the *Session* window as the brief statement:

```
Correlations (Pearson)
Correlation of 4th Yr and 3rd Yr = 0.746, P-Value = 0.000
```

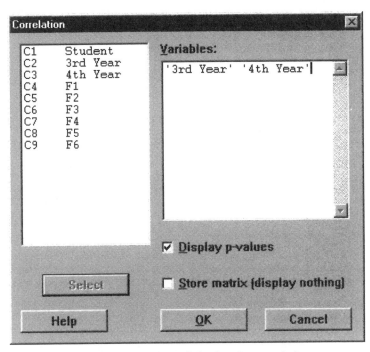

Figure 8.3 MINITAB dialog box for correlation

As is explained in the next section, this shows that there is a *positive* correlation ($r = 0.746$) between students' results in third- and fourth-year and that it is highly significant, with $P = 0.000$ (meaning <0.001).

8.2.3 Interpretation of correlation coefficients

As already noted, correlation coefficients are restricted to values within the range -1.0 through zero to $+1.0$. The interpretation of different coefficients within this range can conveniently be visualized by manipulating the present data to produce sets of fictitious data, purely for illustration. The MINITAB worksheet with the extra columns is in Table 8.2, the details of the manipulations being given briefly in the text.

Figure 8.4 (top), with $r = 1.0$, illustrates a perfect positive correlation. It was obtained by copying and pasting the fourth-year results into a new column labelled *F1* (for fictitious data set No. 1), which was then plotted against the fourth-year results, so that effectively the fourth-year results were plotted against themselves, and all the points fall on a straight line. This is the result that would be obtained if each student in fourth year gained exactly the same mark as in a hypothetical third year where they had marks identical to those in the subsequent fourth year.

Figure 8.4 (middle), with $r = -0.985$, shows an (almost perfect) negative correlation. The small amount of deviation of the points from an underlying smooth line is sufficient to make r not exactly -1.0. The variable *F2* was obtained by rearranging the order of tabulation of the fourth-year marks, to go from the lowest

Table 8.2 MINITAB worksheet with the original (C1–C3) and manipulated (C4–C9) examination marks for the six correlation diagrams in Figures 8.4 and 8.5

C1 Student	C2 3rd Year	C3 4th Year	C4 F1	C5 F2	C6 F3	C7 F4	C8 F5	C9 F6
1	76.8	72.9	72.9	50.9	62.8644	63.3980	60	50.9
2	72.7	70.7	70.7	53.8	57.7721	61.6229	60	53.8
3	75.6	70.0	70.0	55.4	61.4285	65.7167	60	55.4
4	63.9	68.5	68.5	57.8	56.7722	58.9411	60	57.8
5	68.2	67.4	67.4	59.6	46.5160	65.4182	60	59.6
6	62.6	66.2	66.2	59.9	63.8826	64.0844	60	59.9
7	61.2	65.7	65.7	60.3	63.4117	55.5866	60	60.3
8	65.4	64.9	64.9	61.3	54.1699	64.6767	60	61.3
9	60.1	64.2	64.2	62.4	61.4491	46.5152	60	62.4
10	57.6	63.7	63.7	62.8	59.2984	60.7606	60	62.8
11	53.7	63.4	63.4	63.1	67.6389	60.8036	60	63.1
12	65.8	63.1	63.1	63.4	64.9391	59.8994	60	63.1
13	65.2	62.8	62.8	63.7	72.1442	68.2904	60	62.8
14	62.2	62.4	62.4	64.2	67.2785	64.6084	60	62.4
15	61.4	61.3	61.3	64.9	58.2131	61.9158	60	61.3
16	53.1	60.3	60.3	65.7	51.1372	59.4261	60	60.3
17	67.2	59.9	59.9	66.2	61.0460	64.9351	60	59.9
18	50.4	59.6	59.6	67.4	72.2603	55.6929	60	59.6
19	56.9	57.8	57.8	68.5	70.3228	61.4048	60	57.8
20	53.7	55.4	55.4	70.0	58.0196	64.2189	60	55.4
21	51.8	53.8	53.8	70.7	62.0638	67.8544	60	53.8
22	58.7	50.9	50.9	72.9	75.7656	60.8948	60	50.9

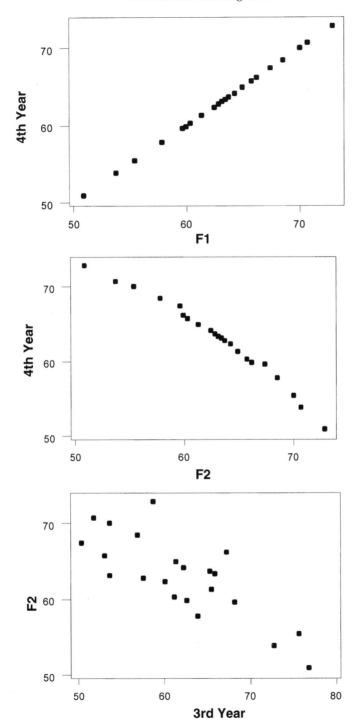

Figure 8.4 Correlation diagrams to illustrate: top, perfect positive correlation, with $r = 1.0$; middle, close to perfect negative correlation, with $r = -0.985$; bottom, negative correlation with $r = -0.791$, similar to the original data, but negative

at the top of the column to the highest at the bottom of the column. This was done with the MINITAB *Sort* command and with *ascending* order as the subcommand. Note that now the *Student No.* in Col. 1 is no longer associated with the entries in *F2* and is simply to be regarded as a convenient set of data for illustrative purposes. The resulting negative correlation of *F2* with fourth-year marks should not therefore be examined for any underlying meaning. However, there are plenty of meaningful negative correlations in other situations. For example, the proportionate incidence of whooping cough (pertussis) in different districts of England and Wales in the late 1970s was negatively correlated with the percentage of infants in these communities who received pertussis vaccine.

Figure 8.4 (bottom), with $r = -0.791$, is another negative correlation, but showing a similar amount of scatter to the original data. It was produced by plotting *F2* against the actual third-year results and, as with the previous plot could well occur with vaccination data but not with student third- and fourth-year exam results.

A correlation coefficient of zero, or close to zero, can occur with different patterns of plotted points, as illustrated in the three sections of Figure 8.5.

Figure 8.5 (top), with $r = -0.074$, is a random scatter of points with an *r*-value close to zero and $P=0.743$, which is not significant. It was produced from the two columns (*F3* and *F4*) of a randomly distributed normal variable with the same mean (62.95) and standard deviation (5.46) as the fourth-year results and then the two columns treated as being third- and fourth-year marks for each student. The method for producing such data is described in FN 3.1.

Figure 8.5 (middle), with $r = 0.0$, shows a horizontal straight line at a *y*-value of 60.0. It was generated by plotting *F5*, which had been filled with 22 entries of the arbitrarily chosen number 60, against the third-year results.

Figure 8.5 (bottom), with $r = 0.077$ and $P = 0.320$, has an inverted letter V-shape. The artificial variable *F6* was produced by copying and pasting the first 10 entries of *F2* (the fourth-year results in ascending order) followed by the last 12 entries of Col. 3, but slightly modified by two entries of 63.1. The resulting plot shows that although the correlation coefficient is close to zero and is non-significant ($P>0.05$), there is nevertheless a definite relationship between the two variables *F2* and *F6*. Graphs of this kind that go up to a peak and come down again, occur in physiological experiments with animals, plants and microorganisms. For example, when a bioactive substance such as a growth factor is given in increasing doses, there may be an increasing response up to a maximum value, similar to the left-hand slope of the figure. At doses beyond a peak critical level there may be diminishing returns as toxicity intervenes and the response falls back to a low value. For the present purposes however, the main message is to highlight the importance of doing a point plot of the data, to reveal any unusual features and before further statistical analysis.

8.2.4 Null hypothesis

Implicit in correlation analysis is the setting up of a null hypothesis (see FN 3.2) which states that 'any apparent relationship between the two variables has arisen by chance, i.e. through random-sampling fluctuations'. The *P*-value that emerges from the statistical processing is the probability of getting the particular value of *r* from

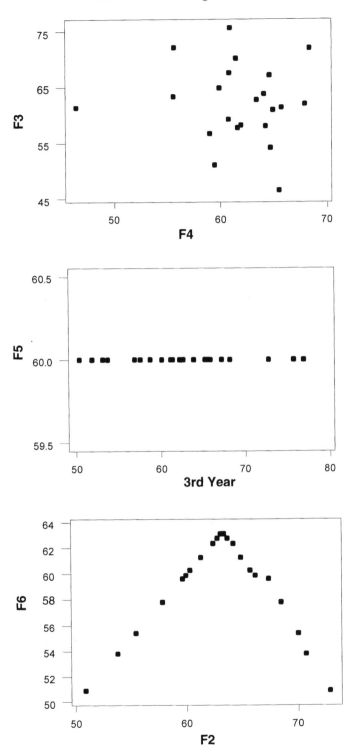

Figure 8.5 Correlation diagrams, all with *r* = zero, or close to zero

the data *if* the null hypothesis is correct. It is never possible to prove that a null hypothesis is correct; the closest is *not to reject it*. By convention, a *P*-value greater than 0.05 is taken as not allowing the null hypothesis to be rejected, with the further conclusion that the *r*-value *could* plausibly have arisen by chance. So although in such a case, with *P* >0.05, there *may* be a relationship between the two variables, the data available do not cross the threshold into the 'significant' zone. In the examples considered so far in this chapter, the correlations have either been very highly significant (Figures 8.1 and 8.4), or firmly in the non-significant zone, with *P* much greater than 0.05.

8.2.5 Correlation and causation

It is important not to assume that because two variables are significantly correlated, that one *caused* the other. There may be a causal relationship, but to establish this requires more than just a significant value of the correlation coefficient. For example, the weight and height of human beings are significantly correlated but that does not mean that height causes weight, or *vice versa*. Rather it is a matter of both variables being expressions of bodily size, which in turn is controlled ('caused') by the processes that regulate growth. This is an area where multiple factors such as age, sex and food intake, as well as height, would be have to be taken into account in order to explain weight. The technique of partial correlation would be applicable and is described in the *MINITAB User's Guide 2*, p. 1–35.

The behavioural and health sciences have provided statisticians with a happy hunting ground for whimsical and nonsensical correlations, particularly when the variables are tabulated by year and for a decade or two. Thus one could probably choose suitable years and locations to show a positive correlation between the number of cases of AIDS and the number of computers in the community; or an inverse correlation between the use of portable telephones and the incidence of bovine spongiform encephalopathy (BSE) in the UK.

On a more serious note (and just to take one example from many), a group of investigators in Finland (Tuomilehto *et al.*, 1990) showed in a survey of different countries that there was a positive correlation between the annual incidence of juvenile diabetes and the annual average consumption of coffee (by adults). The authors were cautious in pointing out that such a positive result merely provided the opportunity to generate a hypothesis, namely that caffeine consumption during pregnancy might be a risk factor for later diabetes in a genetically predisposed foetus. Such a suggestion is biologically plausible from what is known of the pharmacology of caffeine. Thus to progress from correlation to causation, an essential further ingredient is a plausible mechanism. This in turn may lead into the potentially definitive area of the controlled trial or experiment from which stronger evidence of causation might emerge.

8.3 RANK CORRELATION

If either of the variables for correlation analysis is not normally distributed, then the Pearson procedure should be avoided. Instead, a non-parametric approach should be used, with the correlation analysis being based on the *ranks* of the observations. Surprisingly, MINITAB (Release-12) does not provide a program for non-parametric correlation analysis, although the *User's Guide 2* (p. 1–34) briefly mentions Spearman's *rank correlation coefficient* ρ (Greek, *rho*), but without giving the formula or providing the statistical table needed for interpreting the output.

In the previous edition of the present book, the Kendall *rank correlation co-efficient* (*K*) was described and is therefore also given here. Despite this procedure being a standard statistical method, strangely neither MINITAB not Excel provides it, although some other statistics packages do, e.g. *Stat-100* (Biosoft®, 37 Cambridge Place, Cambridge CB2 1NS, UK). However, to avoid introducing a new statistics package at this late stage in the text, what follows is some work with formulae, together with input from MINITAB where possible. Let us hope that later versions of MINITAB will contain the Kendall procedure as a routine item. It may be noted as a reciprocal curiosity that the *Stat-100* package does not provide the Pearson product–moment correlation coefficient.

8.3.1 Preliminaries

As illustrative data for the Kendall procedure it is convenient to use the same sets of third- and fourth-year student examination results as above. The fact that both variables, being normally distributed, are best analysed by the Pearson method does not make Kendall inapplicable. It simply means that Kendall is less suitable because, as a non-parametric method, it is likely to be less powerful at detecting significant effects. For present purposes, however, the same data are being taken largely to avoid having a separate set of results and purely as an example.

Readers who intend to carry out non-parametric correlations beyond the present exercise may wish to use one of the packages that contain the Kendall procedure. In the latter context, note that the third- and fourth-year data columns on the MINITAB worksheet can be copied and pasted into a *Stat-100* Spreadsheet and the Kendall test performed directly thereafter.

In calculating the Kendall procedure manually and with MINITAB, the first step is to select one of the data columns and arrange for the observations to be presented in *ascending* numerical order. Since the primary presentation of the examination results in Table 8.1 gave the fourth-year marks in *descending* mark order, MINITAB must first be used to invert this. Since the whole Kendall procedure needs 12 columns of worksheet, it is convenient to open a new worksheet (as *Worksheet No. 2* within the same MINITAB Project) and copy and paste into it the first three columns of Worksheet No. 1 (Table 8.1). This is shown as the first three columns of Table 8.3.

To list the students, together with their fourth-year marks, in reverse order of merit, requires that the receiving columns in MINITAB be given different titles. Thus Col. 4 is designated *St ID No.*, for *Student Identification Number* and Col. 5 is headed *Sort 4Y*, for *4th Year Marks sorted from the lowest result upwards*. To do the sorting on MINITAB, go to **Manip > Sort** and, in the dialog box enter instructions as in Figure 8.6, i.e.:

---

1) *Sort column(s)*: 4thYear ;
2) *Store sorted column(s) in*: 'Sort 4Y' ;
3) *Sort by column*: Student ✔ Descending ;
4) OK .

Table 8.3 MINITAB worksheet for calculating Kendall's coefficient of rank correlation

C1	C2	C3	C4	C5	C6	C7	C8	C9	C10	C11	C12
Student	3rd Year	4th Year	St ID No	Sort 4Y	Sort 3Y	A	B	S	N	K	Z
1	76.8	72.9	22	50.9	58.7	14	7	127	22	0.5498	3.581
2	72.7	70.7	21	53.8	51.8	19	1				
3	75.6	70.0	20	55.4	53.7	16	2				
4	63.9	68.5	19	57.8	56.9	15	3				
5	68.2	67.4	18	59.6	50.4	17	0				
6	62.6	66.2	17	59.9	67.2	3	12				
7	61.2	65.7	16	60.3	53.1	15	0				
8	65.4	64.9	15	61.3	61.4	10	4				
9	60.1	64.2	14	62.4	62.2	9	4				
10	57.6	63.7	13	62.8	65.2	6	6				
11	53.7	63.4	12	63.1	65.8	4	7				
12	65.8	63.1	11	63.4	53.7	10	0				
13	65.2	62.8	10	63.7	57.6	9	0				
14	62.2	62.4	9	64.2	60.1	8	0				
15	61.4	61.3	8	64.9	65.4	4	3				
16	53.1	60.3	7	65.7	61.2	6	0				
17	67.2	59.9	6	66.2	62.6	5	0				
18	50.4	59.6	5	67.4	68.2	3	1				
19	56.9	57.8	4	68.5	63.9	3	0				
20	53.7	55.4	3	70.0	75.6	1	1				
21	51.8	53.8	2	70.7	72.7	1	0				
22	58.7	50.9	1	72.9	76.8						

See text for the explanation of the column headings.

The same sorting procedure is used on the third-year marks and with the corresponding label *Sort 3Y* in Col. 6.

---

1) *Sort column(s)*: ['3rd Year'];
2) *Store sorted column(s) in*: ['Sort 3Y'];
3) *Sort by column*: [Student] [✔] [Descending];
4) [OK].

8.3.2 Kendall rank correlation coefficient

Having used MINITAB to organize the data into a format suitable for Kendall, the next steps are performed manually. The first step in the Kendall procedure is to obtain entries for the quantity A as listed in Col. 7 of Table 8.3. This is achieved line by line, on the sorted third-year marks in Col. 6 starting from the top. This presupposes that the fourth-year marks are already in ascending rank order in Col. 5.

To find A, take each entry in Col. 6, starting with the topmost and count the number of *larger* values that lie in the column *below* it. For the first entry of 58.7, reading downwards, gives 67.2, 61.4, 62.2, 65.2, 65.8, 60.1, 65.4, 61.2, 62.6, 68.2, 63.9, 75.6, 72.7 and 76.8. This gives a total of 14 in the first line of Col. 7. Similarly, the

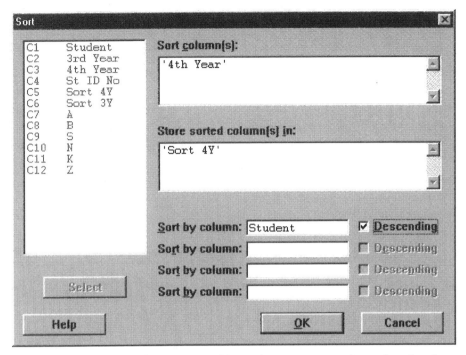

Figure 8.6 MINITAB *Sort* dialog box, with entries to rearrange the student fourth-year marks in descending order of student number (i.e. from 22 at the top of the column to 1 at the bottom). This causes the marks themselves to be in ascending order down the column

next Col. 7 entry of 51.8 has 19 larger values lying below it in the column. Go down the column in this way, entering each *P* value in turn.

To find the *B* values in Col. 8 proceed similarly, except that with each Col. 6 value, count the number of *smaller* Col. 6 numbers that lie below it. For example, the first Col. 6 entry of 58.7, has 51.8, 53.7, 56.9, 50.4, 53.1, 53.7 and 57.6, a total of seven smaller numbers for the first line of the *B* column. This is done line by line down to the last entry in Col. 6, which has nothing below it to be either larger or smaller and therefore there are no entries in the last line of *A* or *B*.

MINITAB can now be used to complete the calculations which involve the following three equations. *S*, the sum of the differences between *A* and *B* is given by:

$$S = \Sigma(A-B) \qquad \text{Eq. 8.1}$$

The Kendall rank correlation coefficient (*K*), which is also known by the Greek letter τ (tau) is:

$$K = 2S/N(N-1) \qquad \text{Eq. 8.2}$$

where *N* is the number of students, or more generally, the number of 'observational units', each of which has two measurements to be correlated. Note that this formula is only applicable if neither data column contains any *tied* values. If ties are present, a more complex version of Eq. 8.2 with correction factors is required. In the present data there is only one tie in one column and so although the more complex Kendall formula should be used, the simpler Eq. 8.2 will give an approximate value that is sufficient for present purposes.

The quantity Z, which is used to determine the *significance* of the correlation, is:

$$Z = \frac{S\sqrt{18}}{\sqrt{[N(N-1)(2N-5)]}}$$ Eq. 8.3

To use these three equations with MINITAB, it is convenient to label four extra columns on the worksheet to display $N = 22$ and to provide destination columns for the output from the calculations. For Eq. 8.1, go to **Calc > Calculator** and, in the dialog box, enter:

1) *Store result in variable:* $\boxed{S}$;
2) *Expression:* $\boxed{\text{SUM(A-B)}}$;
3) $\boxed{OK}$.

This gives 127 as the calculated value of S. The command *SUM* is the MINITAB command for Σ.

For Eq. 8.2 the corresponding entries are:

1) *Store result in variable:* $\boxed{K}$;
2) *Expression:* $\boxed{\text{2*S/N/(N-1)}}$;
3) $\boxed{OK}$.

Which gives 0.55 (after rounding) as the calculated value of K. The asterisk (*) in the expression indicates multiplication and the slash (/) indicates division.

For Eq. 8.3, the entries are:

1) *Store result in variable:* $\boxed{Z}$;
2) *Expression:* $\boxed{\text{S*SQRT(18)/SQRT(N*(N-1)*(2*N+5))}}$;
3) $\boxed{OK}$.

This gives 3.58 (after rounding) as the value of Z, the *SQRT* being the MINITAB command for $\sqrt{}$.

The quantity Z is the so-called *standard normal deviate* and is used to determine the probability value (P) of the Kendall coefficient. If $Z < 1.96$, the correlation is *not* significant ($P > 5\%$); if $Z \geq 1.96$, the correlation is *significant* ($P < 5\%$), and if $Z \geq 2.58$, the correlation is *highly* significant. In the present example, a Z of 3.58 is *very* highly significant. Note that the Kendall coefficient resembles the Pearson coefficient in having values only between -1.0 and $+1.0$, with zero representing no correlation and for the same reasons as in Figure 8.5. Likewise the null hypothesis is as for Pearson.

The output from Biosoft's *Stat-100* package is $K = 0.555$ and $Z = 3.617$ which are both within 1% of the above.

8.3.3 Procedural refinements with Kendall

The calculation of the Kendall rank correlation coefficient becomes more complicated if either column of results contains *ties*, and if N is 10 or less. If ties *are* present, there is a need for correction factors in the calculation of K. With the data under consideration, there are no ties in the fourth-year results but there is only one tie in the third-year results. This has only a trivial effect and is scarcely worth correcting for. However, if data sets contain substantial numbers of ties or if N is 10 or less, then specialist advice should be sought.

8.3.4 Spearman rank correlation coefficient

Biosoft's *Stat-100* program offers a second, non-parametric test for correlation, namely that of Spearman, which is based on the *ranks* of the data in each column of marks. With the third- and fourth-year student results under consideration, *Stat-100* gives a Spearman rank correlation coefficient of 0.713, with a *P*-value of 0.000, i.e. <0.1%, and therefore highly significant.

8.4 LINEAR REGRESSION

In contrast to its rather scanty provision for *correlation*, MINITAB is rich in procedures for *regression* analysis. Indeed the problem for the beginner is to choose from the extensive and branching menus, since MINITAB offers such a diversity of possibilities and choices. Even the shortest request for *regression* is answered by a generous outpouring of information, not all of which may be wanted in the first instance. Therefore this section starts in a basic fashion and then moves on to the extra insights, refinements and controls.

8.4.1 Preliminaries for regression

The same data set will be used as for the correlation analysis, partly for continuity but more particularly as an example of the common case where the data consist of *measurements*, as distinct from *counts* or *proportions*.

It is convenient to open a new worksheet, *Worksheet No. 3*, within the already existing MINITAB project (*Chap 8. MINITAB.mpj*), and then copy and paste in the first three columns with student identification (ID) number and the third- and fourth-year marks, as in Table 8.1. Regression analysis requires that each observation unit (student) should be independent of each other (e.g. no cheating in the exams!) and have two observations associated with it. If one observation is missing, the other will not be used by MINITAB.

One of the columns of data has to be selected as the *independent variable*, what MINITAB calls the *predictor*. This will be plotted on the *x*-axis, or abscissa, while the other variable is on the *y*-axis, or ordinate. This latter is the *dependent* variable, referred to by MINITAB as the *response*. The underlying principle in the regression analysis is that the independent variable is reckoned to influence the dependent variable, but not vice versa. In the present example it is logical to take the third-year mark as the independent, or predictor, variable, and the fourth-year mark as the dependent, or response, variable.

Another feature of regression is that the error variation is all assigned to the response variable. So, when error bars or confidence intervals are added to the graph, they refer to variation in the response and not to fluctuations in the independent variable.

8.4.2 Fitting the best straight line

Fitting the best straight line is achieved by the procedure known as the *method of least squares*. Using MINITAB with the same points that have already been plotted (Figure 8.1) for correlation analysis, go to **Stat > Regression > Fitted Line Plot** and, when the dialog box opens, fill the spaces as in Figure 8.7:

1) *Response [Y]:* ['4thYear'];
2) *Predictor [X]:* ['3rd Year'];
3) *Type of Regression Model*
   ⬤ *Linear*;
4) Open the [Options] and [Storage] sub-windows and check that no additional output is requested meanwhile; then return to the main window;
5) [OK].

This gives Figure 8.8 as the plot of fourth-year versus third-year student marks, and with the best fitting straight line inserted on top. It looks as if there is an approximately linear relationship between the two variables but with much scatter.

The straight line may also be described as the *regression* of fourth-year marks on

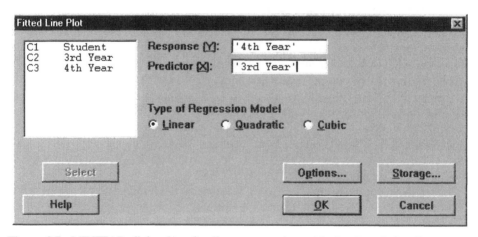

Figure 8.7 MINITAB dialog box for linear regression with the best-fitting line superimposed on the points

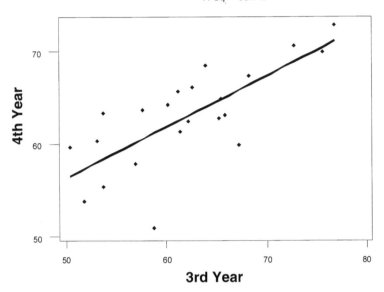

Figure 8.8 Linear regression of student fourth-year marks on their third-year marks

those from third-year. The word itself was introduced by the English anthropologist, Sir Francis Galton (1822–1911), who collected and analysed data on the heights of children in relation to those of their parents. This was one of the earliest investigations in scientific statistics. He showed that tall parents tended to have tall children, and vice versa. However, the heights of the tall children, instead of equalling or excelling those of their parents, tended to be smaller and to be closer to the overall average height of children. He called this phenomenon *regression towards the mean,* from which the word *regression* for such relationships has persisted to the present day.

8.4.3 Primary statistics of regression

When entered with data as above, MINITAB not only produces the graph in Figure 8.8, but also provides additional statistical output in the *Session* window, as recorded in Table 8.4. This has been sectionalized to assist discussion, which is best done in the sequence A, D, C, B, and not as MINITAB gives it.

Section A: gives the regression equation as:

$$y = 28.7 + 0.553x \qquad \text{Eq. 8.4}$$

which can be generalized as:

$$y = a + bx \qquad \text{Eq. 8.5}$$

where b is the slope of the line, and a is the intercept on the y-axis when $x = 0$.

Table 8.4   Output in the *Session* window of MINITAB that accompanied the graphical plot
in Figure 8.8. To assist discussion in the text, the output has been sectionalized A–D

```
Section A
Regression
The regression equation is
y = 28.7 + 0.553 x

Section B
Predictor            Coef          St Dev        T           P
Constant             28.661        6.890         4.16        0.000
X                    0.5530        0.1104        5.01        0.000

Section C
S = 3.728            R-Sq = 55.7%              R-Sq(adj) = 53.4%

Section D
Analysis of Variance

Source               DF            SS            MS          F         P
Regression           1             348.78        348.78      25.10     0.000
Residual Error       20            277.92        13.90
Total                21            626.70
```

Putting Eq. 8.4 into ordinary language, and assuming the validity of extrapolation
(which is often, as here, unwise!), the *a* term means that a student who had $x = 0\%$
in the third year would be expected to get 28.7% in the fourth year. This ignores the
non-statistical fact that such a student would probably not be admitted to fourth
year! However, in the mark range 50–77% of the actual observations, a student
with 50% in the third year could be expected to get $28.7+0.553\times50 = 56\%$ in the
fourth year. The slope term *b*, of 0.553 means that third-year marks above 50%
translate into fourth-year marks above 56%, at a rate of just over one-half mark-
per-mark. So while a third-year student with 50+0 can expect 56% in the fourth
year, a third-year student with 60 (i.e. 50+10) could expect only $56.35+0.553\times10 =$
61.9% in the fourth year. And a third-year student with 50+20=70 marks would
only expect $56.35+0.553\times20 = 67.4$ marks in the fourth year. So, to get 70% in the
fourth year, the third-year student would expect to need $(70-28.7)/0.553 = 74.7\%$.
Finally, to achieve 100% in the fourth year would need theoretically 129% in the
third year, which again highlights the perils of extrapolating an equation beyond
the range of the actual observations. In other words, the linear regression is only
approximately linear and within the limits explored.

    Section D of Table 8.4 is an *analysis of variance* (§3.2) of the regression. With *N*
= 22 students, there are 21 degrees of freedom, as shown opposite *Total* in the *DF*
column. These 21 *DF* are subdivided into 1 for the regression line itself and 20 for
Error or *Residual* variation. The sums of squares (*SS*) and mean square (*MS*) are as
previously described for analysis of variance (§3.2.4). The variance ratio (*F*) is the
Regression MS divided by the *Error MS*, and its high value of 25.10 is associated
with a very low value of *P*, the probability associated with the null hypothesis. This
latter states that any apparent regression, i.e. *slope different from horizontal (zero)*
is due to random-sampling fluctuations, and not a 'real' slope. The low value of *P*

means that the null hypothesis can be dismissed, and the regression therefore stated as being *highly significant*.

Section C contains two separate pieces of information. The $S = 3.728$ is the *error standard deviation*, namely the standard deviation of the distances of the points vertically below and above the fitted line. It is the square root of the *Residual Error MS* = 13.90 given as part of the analysis of variance in Section D.

The *R-Sq* = 55.7% is the percentage of the variability in the *y* variable that is accounted for by the regression, the rest of the variability being due to 'error' or unaccountable scatter. The *R-Sq(adj)* is a slightly more refined estimate of the same quantity that appears as r^2 in formulae. It is calculated from the *SS* figures in the analysis of variance in Section D, and is the ratio of the regression SS/total SS, i.e. 348.78/626.70 = 0.556 53, which converts to 55.7%. Its formal name is the *coefficient of determination*.

Section B contains an analysis of the significance of the $a = 28.7$ and $b = 0.553$ terms in the linear equations (Eqs 8.4 and 8.5). Each is examined, by the Student's t-test, for significance of difference from zero and the output presented as T (for Student's t) and P for probability value. The very *low* values of P (for the null hypothesis) means that there is a *high* probability of neither a nor b being zero.

8.5 ADDITIONAL STATISTICS OF LINEAR REGRESSION

What has been done above may be described as some of the *primary statistics of regression*. MINITAB offers a lot more. Additional questions to be explored include:

- What confidence intervals should be attached to the predicted fourth-year marks?
- How do we know the regression is linear?

8.5.1 Prediction intervals (graphical)

Return to **Stat > Regression > Fitted Line Plot** and, when the dialog box appears it should still contain the entries made previously (Figure 8.7). Now the *Options* button will be used in step 4 below:

---

1) *Response [Y]:* 4thYear ;
2) *Predictor [X]:* '3rd Year' ;
3) *Type of Regression Model*
   ● *Linear*;
4) Open the Options sub-dialog box and at *Display Options* (Figure 8.9), select:
   ☑ *Display confidence bands*
   ☑ *Display prediction bands*
5) Click on OK to close the sub-dialog box and then again to close the main dialog box.

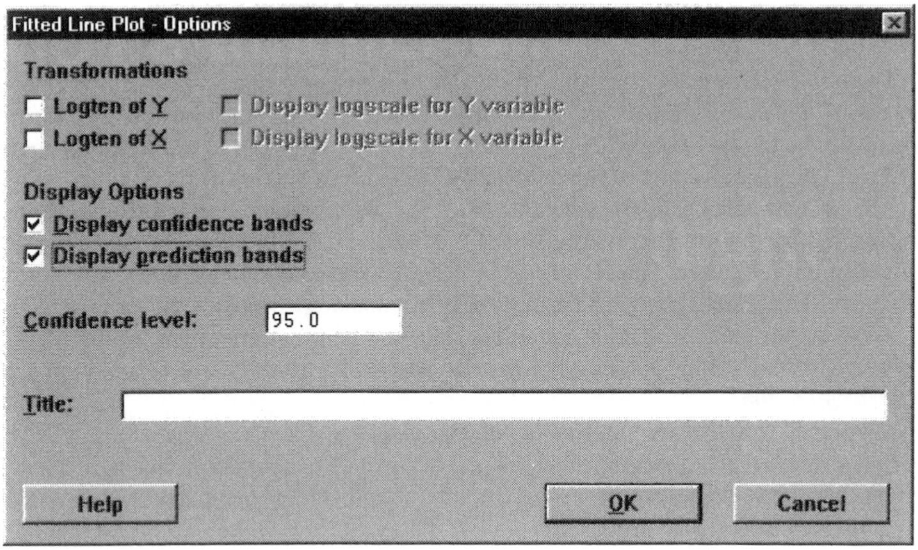

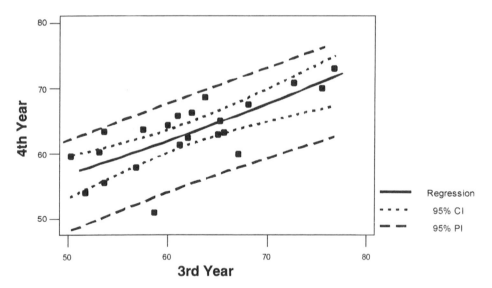

Figure 8.9   Sub-dialog box with entries for selecting confidence and prediction bands

Figure 8.10   Regression of student fourth-year marks on their third-year marks, with 95% confidence bands (dotted lines) for the regression, and 95% prediction bands (dashed lines)

Figure 8.10 shows the regression diagram surrounded by additional dotted and dashed lines (with some artwork adjustments from how MINITAB delivered it). The inner, dotted, lines are the 95% confidence bands of the regression itself. These are narrowest at the centre, indicating that the position of the regression line is known with greatest certainty in the middle region, and with somewhat wider uncertainty as the dotted lines flare out slightly at the ends.

Table 8.5 Approximate *point estimates*, and 95% *prediction intervals*, for student fourth-year marks from their third-year marks, as read from the regression diagram (Figure 8.10)

3rd Year mark	Point estimate of expected 4th year mark	95% Prediction intervals of expected 4th year mark
50	56	47, 63
55	59	52, 66
60	62	54, 68
65	65	56, 72
70	67	59, 75
75	71	62, 77

Of greater interest are the outer, dashed lines that define the *95% prediction bands*. These allow a prediction of the likely range of fourth-year marks to be made from chosen third-year marks. While the regression line itself allows *point estimates*, as in §8.4.3 above, the 95% *prediction bands* attach a probability statement. Thus reading approximate values by eye from the graph, the output is as in Table 8.5.

Note that a mark of 65% in the third year translates into the same expected mark in the fourth year, but with wide confidence intervals, from 56 to 72.

8.5.2 Prediction intervals (exact)

More exact prediction intervals can be obtained as a printout from MINITAB by following a different path through the *Regression* menus. First, however, the worksheet needs a set of third-year marks from which fourth-year point estimates and prediction intervals are wanted. This has been done (Table 8.6) in Col. 4, labelled *3Y Input* in Worksheet No. 3 from which the regression calculations are made. Table 8.6 also contains the output from MINITAB, generated from the following: Go to **Stat** > **Regression** > **Regression** and, when the dialog box appears (Figure 8.11) enter:

1) *Response:* '4thYear' ;
2) *Predictors:* '3rd Year' ;
3) Click on Options and, when the sub-dialog box (Figure 8.12) opens,
4) Enter '3Y Input' under *Prediction intervals for new observations:*
5) Check ✔ the boxes opposite *Fit intercept, Fits* and *Prediction limits*, as shown in Figure 8.12;
6) Click on OK to close the sub-dialog box and then again to close the main dialog box.

The output appears in the next three empty columns of the worksheet, as shown in Table 8.6. Col. 5 has the more accurate values of the point estimates (*'fits'*, see below) for the expected fourth-year marks, from chosen third-year marks of 50, 55, 60, 65, 70 and 75%. In Cols. 6 and 7 are the more exact values of the 95% prediction intervals than those read previously off the graph (Table 8.5).

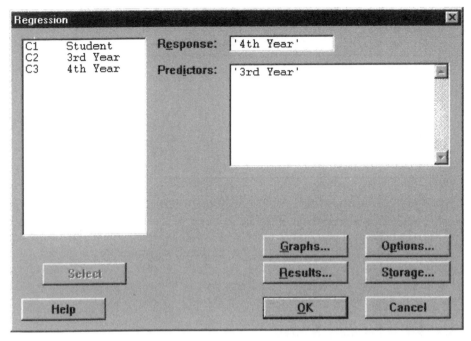

Figure 8.11   MINITAB dialog box with entries for doing the regression of student fourth-year marks on their third-year marks

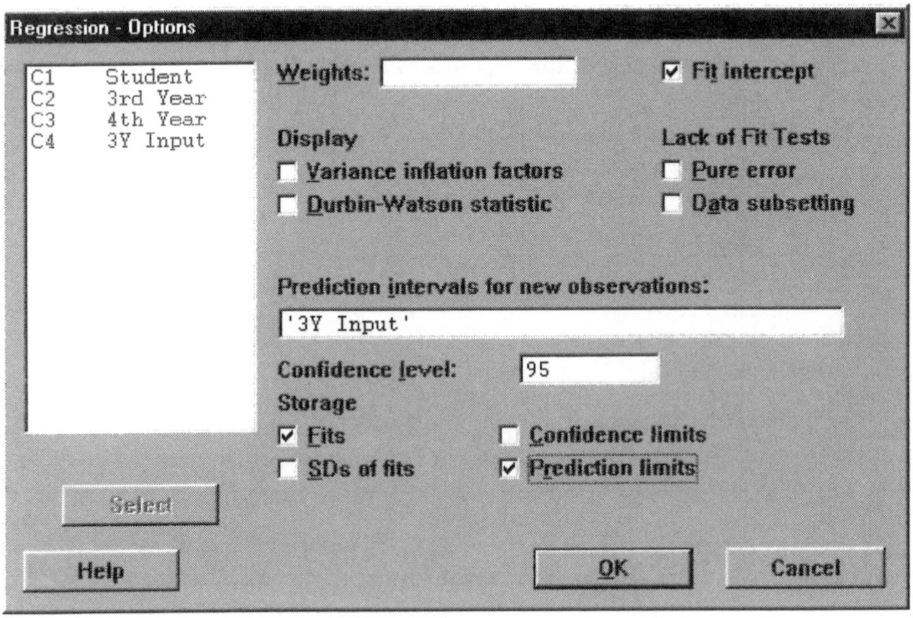

Figure 8.12   MINITAB sub-dialog box for point estimates and confidence intervals of fourth-year marks from selected third-year marks

Table 8.6 Extension of MINITAB Worksheet (No. 3) with additional columns for getting exact point estimates (PFIT1) and 95% prediction intervals (PLIM1, PLIM2) of student fourth-year marks from chosen third-year marks (3Y input)

C4 3Y Input Chosen 3rd- year mark	C5 PFIT1 Point estimate of 4th-year mark	C6 PLIM1 95% Prediction Interval Lower	C7 PLIM2 Upper
50	56.3094	47.8918	64.7271
55	59.0742	50.9615	67.1870
60	61.8391	53.8750	69.8031
65	64.6039	56.6235	72.5842
70	67.3687	59.2080	75.5294
75	70.1335	61.6389	78.6280

8.5.3 Is the regression necessarily linear?

It has been assumed so far that the regression is linear, mainly because this is the simplest assumption and there is no other information to suggest a different relationship between third- and fourth-year marks. However, the relationship between the two variables might be better represented by a curve. This can be explored (among numerous other possibilities) by returning to §8.5.1 and Figure 8.7, and selecting the $\boxed{\bullet}$ *Quadratic* option, together with the options for 95% confidence bands and 95% prediction bands as before. This gives Figure 8.13 as the graphical output and Table 8.7 as the results in the Session window. Graphically, the degree of curvature is only slight, because the quadratic equation (Eq. 8.6) in

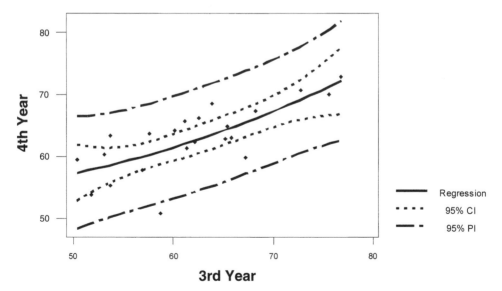

Figure 8.13 Quadratic regression of student fourth-year marks on their third-year marks, with 95% confidence bands for the regression (dotted lines), and 95% prediction bands (dashed lines)

Table 8.7   Output, in the *Session* window of MINITAB, for quadratic regression of the third and fourth year student marks. The output is sectionalized as was that of the *linear* regression in Table 8.4, and the items for special attention placed in ⎧boxes⎫

```
Section A
Polynomial Regression

Y = 60.7112 - 0.474876X + 8.13E-03X**2
```

i.e. $\boxed{y\ =\ 60.71\ -\ 0.475x\ +\ 0.00813x^2}$ Eq. 8.6

```
Section B
(No entry corresponding to that in Table 8.5)

Section C
R-Sq = 56.5%

Section D
Analysis of Variance
```

SOURCE	DF	SS	MS	F	P
Regression	2	353.940	176.970	12.3277	3.70E-04
Error	19	272.755	14.356		
Total	21	626.695			

SOURCE	DF	Seq SS	F	P
Linear	1	348.780	25.0997	6.72E-05
Quadratic	1	5.160	0.359472	$\boxed{0.555883}$

Table 8.7 contains a relatively small x^2 term. A major difference from the linear equation is in the intercept $a = 60.71$, which would mean that a student with $x = 0\%$ in the third year could expect 60.71% in the fourth year, which is even less likely than the 28.7% predicted by the linear regression. All this goes to show that with a single data set of only $N = 22$ pairs of observations, there are limits to the statistical insights afforded, and long extrapolation beyond the observations should be avoided. Maybe if there were numerous sets of third- and fourth-year marks, a clearer picture of the linearity or otherwise of the regression would emerge.

A more objective reason for discarding the quadratic regression is provided by the analysis of variance in Section D of Table 8.7. The boxed $\boxed{P = 0.556}$ associated with the quadratic element is not significant, which means that there is no supportive evidence for an x^2 in the equation. Thus, although the regression may well have some deviation from linearity, the information on hand does not allow the nature of the deviation to be determined. Likewise the ⎡●⎤ *Cubic* regression option in Figure 8.7 does not support the introduction of an additional x^3 term. Of course, the quadratic and cubic choices do not exhaust the mathematical possibilities and, for example, a logarithmic transformation of the third- and/or fourth-year marks might be beneficial.

8.5.4 Fits and residuals

There are two additional terms, *fits* and *residuals*, that are needed when discussing a regression line and which can be used to explore further aspects of the validity of the 'underlying model' (i.e. that the two variables are truly linearly related). The

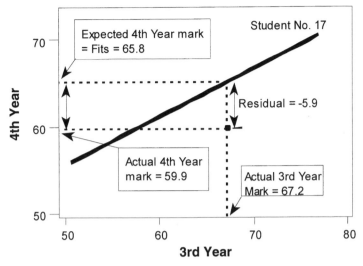

Figure 8.14 Linear regression with additions, to illustrate the meaning of *Fits* and *Residuals*

meaning of *fits* and *residuals* is explained by illustration in Figure 8.14, which shows the linear regression plot as before, but with all the points, except one (Student No. 17), removed for clarity. This student had a third-year mark of 67.2%, followed by 59.9% in the fourth year. As shown in the diagram, the fourth-year mark was considerably lower than the student would have expected, based on the regression line. Extrapolating from the student's plotted point up to the regression line, and then across to the fourth-year axis, shows that the *expectation* was for that student to have received 65.8%. This latter figure is the *fits* mark, since it describes the *expected* fourth-year mark, obtained by *fitting* the third-year mark onto the regression line. The *residual*, indicated by the double arrows in Figure 8.14, is the *difference* between the *expected* and the *actual* fourth-year marks, in this case –5.9. Useful information can emerge from a more detailed examination of the fits and residuals of all the points.

For MINITAB to generate, store and plot various features of fits and residuals, go to **Stat** > **Regression** > **Regression** and, when the dialog box appears (Figure 8.11) enter:

1) *Response:* '4thYear';
2) *Predictors:* '3rd Year';
3) Click on [*Storage*] and, when the sub-dialog box (Figure 8.15) opens,
4) Check the boxes opposite:
   ✔ *Fits*
   ✔ *Residuals*;
5) Click on [*OK*] to close the sub-dialog box and then again to close the main dialog box.

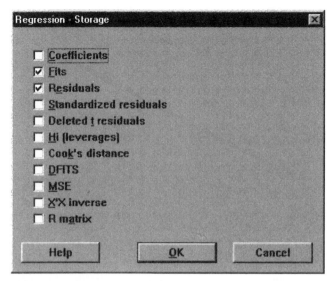

Figure 8.15   MINITAB sub-dialog box of *Regression*, with entries to obtain *Fits* and *Residuals*

The worksheet (Table 8.8) now shows that MINITAB has entered, in the next available columns, the *fits* (FITS3) and the *residuals* (RESI3) opposite each student. Inspection of the output for Student No. 17 will confirm (with extra decimals) the figures used as illustration in Figure 8.14, namely a *fits* value of 65.8 and a *residual* of –5.9.

8.5.5   Uses of fits and residuals

In statistical language, the assumed linear relationship between third- and fourth-year marks is the *model* (of underlying reality). *Fits* and *residuals* can be used to examine the goodness-of-fit of the actual data to the model, therefore a kind of quality control. MINITAB offers four standard plots for this purpose that can be accessed as the sub-dialog box *Graphs* in Figure 8.11. Therefore go to **Stat> Regression > Regression** and, when the dialog box appears (Figure 8.11), confirm that the front entries are:

1) *Response:* '4thYear';
2) *Predictors:* '3rd Year';
3) Click on *Graphs* and, when the sub-dialog box (Figure 8.16) opens,
4) Check the boxes opposite:
   ✔ *Histogram of residuals*
   ✔ *Normal plot of residuals*
   ✔ *Residuals versus fits*
   ✔ *Residuals versus order*
5) Click on *OK* to close the sub-dialog box and then again to close the main dialog box.

Table 8.8 Section of MINITAB worksheet with *Fits* (FITS3) and *Residuals* (RESI3) from the regression of student fourth-year marks on third-year marks

C1 Student	C2 3rd Year	C3 4th Year	C8 FITS3	C9 RESI3
1	76.8	72.9	71.1288	1.7712
2	72.7	70.7	68.8617	1.8383
3	75.6	70.0	70.4652	–0.4652
4	63.9	68.5	63.9956	4.5044
5	68.2	67.4	66.3733	1.0267
6	62.6	66.2	63.2767	2.9233
7	61.2	65.7	62.5026	3.1974
8	65.4	64.9	64.8250	0.0750
9	60.1	64.2	61.8943	2.3057
10	57.6	63.7	60.5119	3.1881
11	53.7	63.4	58.3554	5.0446
12	65.8	63.1	65.0462	–1.9462
13	65.2	62.8	64.7144	–1.9144
14	62.2	62.4	63.0556	–0.6556
15	61.4	61.3	62.6132	–1.3132
16	53.1	60.3	58.0236	2.2764
17	67.2	59.9	65.8204	–5.9204
18	50.4	59.6	56.5306	3.0694
19	56.9	57.8	60.1249	–2.3249
20	53.7	55.4	58.3554	–2.9554
21	51.8	53.8	57.3048	–3.5048
22	58.7	50.9	61.1202	–10.2202

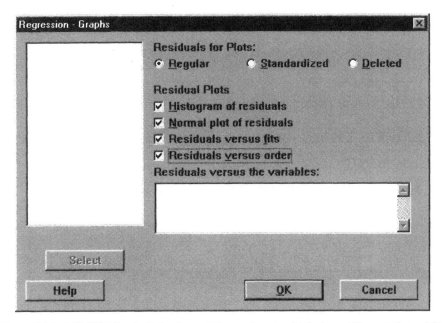

Figure 8.16 MINITAB sub-dialog box of *Regression*, to obtain four different plots of *Fits* and *Residuals*

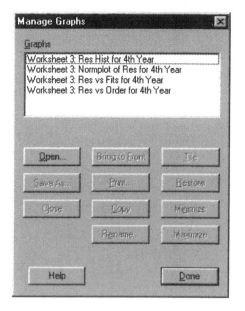

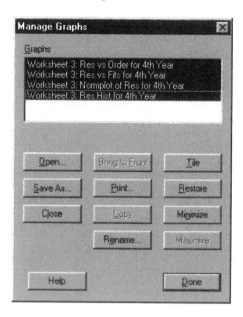

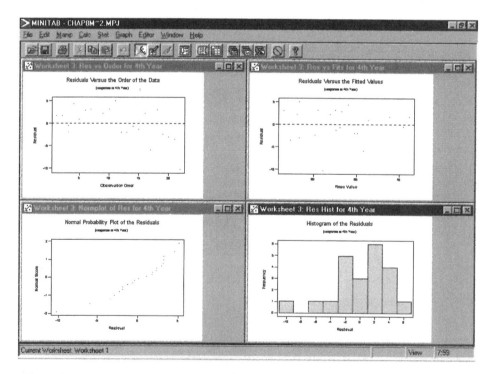

Figure 8.17   Manipulating the *Manage Graphs* dialog box to obtain a *tiled* presentation of the four residuals plots (see text)

The output is four graphs that flash onto the screen in rapid succession but which can be arranged to appear simultaneously (known as *Tiling*) by going to **Windows > Manage Graphs**, whereupon a dialog box will appear (Figure 8.17, top left). In it will be listed all the graphs generated so far in the MINITAB Chapter 8 project (which is a lot!). So, in fact it is useful to have closed beforehand all the previous graphs before generating the new set of four for analysis of *fits* and *residuals*. This was done to produce the shown appearance of the *Manage Graphs* dialog box. To proceed, the list of graphs is highlighted as in Figure 8.17 (top right) and then the *Tile* button on the *Manage Graphs* dialog box is clicked. This arranges the four graphs on the screen together as in the lower part of Figure 8.17. To assist discussion, the diagrams have been subjected to artwork (enlargement of points and legends) and rearranged to give Figure 8.18.

If the original third- and fourth-year marks are truly related by a linear regression 'model', and without complications, then the four graphs of Figure 8.18 should show the following:

- **Residuals versus the fitted values ('fits')**, in the top left: should have a random pattern of points on either side of the Residual = zero line. The result here is satisfactory.
- **Normal probability plot of the residuals**, in the top right: should lie on a straight line if the residuals are normally distributed. Departures from linearity may indicate that some of the assumptions of the regression 'model' are not valid. The present result, although approximately linear, has slight downward curvature. It would be useful to have data sets from other groups of students in the same courses to see if this is a reproducible observation. If so, expert advice should be sought.
- **Histogram of the residuals**, in the lower left, should approximate to a normal (bell-shaped) distribution with a mean of zero. Here there is some suggestion of skewness and with the centre to the right of zero. Again, further data sets would be desirable to see if this is a reproducible phenomenon.
- **Residuals versus the order of the data**, in the bottom right, should have a random scattering of points around a residual value of zero. The present data show some downward trend from left to right, with the early observations being above the line and the later ones below it. This indicates that a source of non-random error may be present, which if repeated in other data sets should be investigated further.

8.6 SUMMARY

This chapter has followed the general pattern in this book of taking a single set of observations and using them to illustrate a varied menu of statistical attitudes and procedures. Here the data set consisted of student examination marks at the end of the third and fourth years at university – not exactly 'experimental biology' – but illustrative of many types of measurement data that are suitable for correlation and regression analysis. With correlation, both parametric and non-parametric methods were used, and emphasis was given to the distinction between correlation and *causation*. With regression, both linear and higher order regressions were explored,

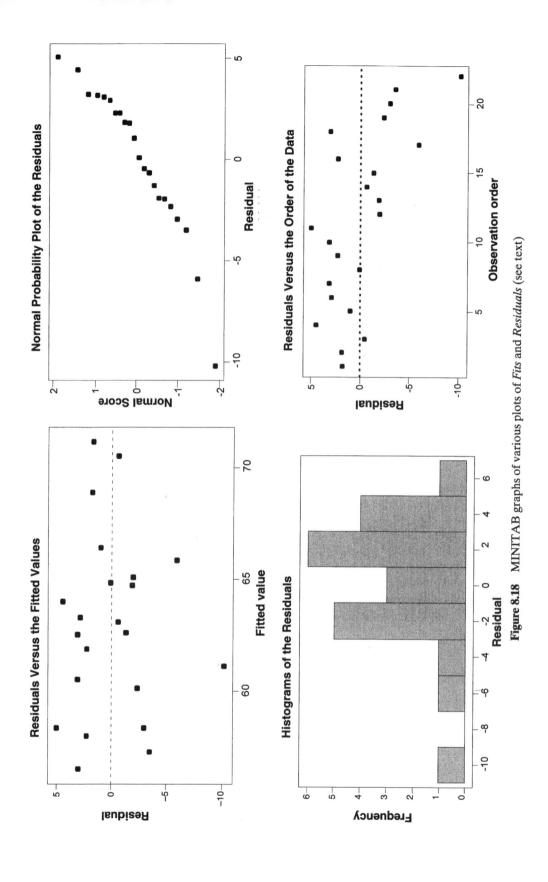

Figure 8.18 MINITAB graphs of various plots of *Fits* and *Residuals* (see text)

and methods defined for calculating intercepts and confidence intervals. Finally, the quality of the underlying regression 'model' was examined through various plots of *residuals* and *fits*. The material presented is no more than an introduction to the subject. For example, partial correlation was not touched on, nor were the several sophisticated regression options offered by MINITAB investigated. The next chapter, however, deals with a special application of regression, namely standard curves and dose–response assays.

9

Dose–Response Lines and Assays

When I was a child, I spake as a child, I understood as a child, I thought as a child; but when I became a man, I put away childish things.

The First Epistle of Paul to the Corinthians (xiii, 2)

9.1 DOSE–RESPONSE INTERPOLATION ASSAY

9.1.1 Scope and generalities

In the biuret method for assaying protein, a dose–response graph is prepared with a standard protein solution, such as bovine serum albumin (BSA). Table 9.1 is a typical set of results, from which Figure 9.1 has been plotted as the standard dose–response line. Protein solutions of unknown concentration were tested in parallel and, from their optical absorbancies, their protein contents can be estimated. Rough estimates can be obtained by eye, directly from the graph, but a full evaluation requires calculation. What follows is an account of the procedures for estimating the protein content of unknowns, with 95% confidence intervals for the estimates. Included are methods for checking the validity of the assay. Both regression analysis and analysis of variance are involved and the reader is assumed to have seen the previous chapters on these subjects. Although MINITAB is used throughout, there is no integrated procedure in the MINITAB portfolio specifically provided for this type of assay, therefore some formula work with the MINITAB *Calculator* function is required.

Table 9.1 Biuret test*: Optical absorbancies (555 nm, 1 cm) given by different amounts of a standard protein and by three unknown samples

Sample	Amount used	Absorbancies in duplicate tests
Standard protein	1.0 mg	0.056, 0.080
	2.0 mg	0.122, 0.138
	4.0 mg	0.256, 0.278
	6.0 mg	0.400, 0.394
	8.0 mg	0.531, 0.543
Unknown A	1.0 ml	0.051, 0.041
Unknown B	1.0 ml	0.274, 0.286
Unknown C	1.0 ml	0.497, 0.506

*Method of Herbert, D., Phipps, P.J. and Strange, R.E. (1971) Chemical analysis of microbial cells, pp. 209–344, in: J.R. Norris and D.W. Ribbons (Eds.), *Methods in Microbiology, Vol. 5B.* Academic Press, London.

The biuret test (Herbert *et al.*, 1971), although somewhat old-fashioned, was chosen because it has a convincing linear relationship between dose and response, without complications that occur with some other assays, e.g.

- Top-end curvature.
- Systematic changes in standard deviation along the dose/response line.
- The need to transform one of the variables.

It is given here as a general example of an assay where the response is a *measurement*, such as optical absorbance, as distinct from a *proportion*. Assays involving the latter are discussed in §9.3 below, in the context of determining ED_{50} and LD_{50}. Other common types of measurement responses are zone diameters of growth or inhibition on microbial culture plates, radioactivity counts, and a wide variety of physiological or morphological parameters of animals and plants. Count data, if the

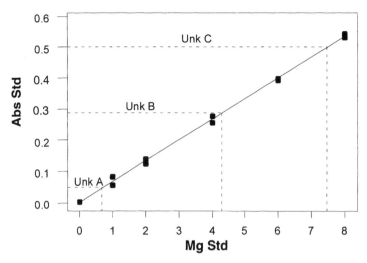

Figure 9.1   Dose–response line of the biuret assay for protein, fitted through the origin, with
results for three unknowns estimated graphically

counts are large enough, can be treated as measurements (see Chapter 6), in
dose–response experiments such as heat-killing of microorganisms, to determine
the decimal-reduction time (D-value).

9.1.2   MINITAB worksheet

The information in Table 9.1 can not be used by MINITAB in that format. Instead
the data should be presented as in the first four columns of Table 9.2. The first two
columns contain the dose (C1, *mg Std*) and response (C2, *Abs Std*) for the Standard
protein, and Cols 3 and 4 the corresponding information for the Unknowns. With
the latter, the dose in mg is unknown and the purpose of the assay is to estimate it.
Therefore the Unknowns were all tested in an equal volume of 1.0 ml and identified
by number (C3, *Unk No*), where 1 = Unknown A, 2 = Unknown B and 3 =
Unknown C. The optical absorbancies (C4, *Unk Abs*) in duplicate tests are given

Table 9.2   MINITAB worksheet for standard curve assay (see text for explanation of
columns)

C1 mg Std	C2 Abs Std	C3 Unk No	C4 UnkAbs	C5 U No	C6 Umean A	C7 U mg
1	0.056	1	0.051	1	0.0460	0.71164
1	0.080	1	0.041	2	0.2800	4.20418
2	0.122	2	0.274	3	0.5015	7.51015
2	0.138	2	0.286			
4	0.256	3	0.497			
4	0.278	3	0.506			
6	0.400					
6	0.394					
8	0.531					
8	0.543					

alongside. Thus C1–C4 on the worksheet are to be read as two separate tables, with C1 and C2 having the dose–response information for the Standard, and C3 and C4 the identity and the responses of the Unknowns.

C5 also contains the *identities* of the Unknowns, but with just a single entry for each. It has been given a different name (*U No*), since MINITAB does not allow two columns on a worksheet to have the same name. C6 contains the mean absorbancy (*Umean A*) for each unknown and was obtained from **Stat** > **Descriptive Statistics**, as described in §1.8, and copied into the worksheet from the Session window.

The final column (C7, *U mg*) has the estimated mg in 1 ml of each Unknown and was obtained from the Standard dose–response curve (C1, C2) as described below. Thus the first seven columns of the worksheet are to be read as three separate tables: C1and C2 for the Standard, C3 and C4 for the Unknowns; and C5–C7 for the means of the Unknowns.

9.1.3 Graphical evaluation

It is necessary here to backtrack slightly. As a generality with numerical results, it is useful to have a graphical plot of the data before the statistical analysis. In the present instance, the 'pre-statistical' plot should show the dose–response line going through the origin at $x = 0$ and $y = 0$. This is because the biuret test also contained a *blank* sample, not reported in Table 9.1 and which had no protein in it, but just the biuret reagent plus diluent. This sample was used to set the spectrophotometer to zero absorbance. Therefore the plotted dose–response line should be shown as going through zero on both the *x*- and *y*-axes. To do this on MINITAB, it is necessary to add, temporarily, extra data to C1 and C2 on the worksheet. In C1, on the line under the second '8' is inserted a 0 (zero) which likewise is also inserted under the 0.543 in C2. Then a point plot of Cols 1 and 2, followed by some additional artwork, gives Figure 9.1. The additional artwork included inserting the straight line manually, and also interpolating the mean absorbancies of the Unknowns to gain approximate estimates of their protein content. This gives rough estimates of protein, in mg/ml as:

Unknown A: 0.7
Unknown B: 4.3
Unknown C: 7.4

These rough estimates are close to the exact estimates already presented in Table 9.2 and which emerged from the regression analysis, below. Irrespective of subsequent calculations, it is worthwhile to have assurance that the dose–response relationship is indeed a straight line, and rough estimates of the results, in case errors occur in the more exact procedures.

Finally, having obtained Figure 9.1, the two entries of zero in C1 and C2 should be deleted to avoid spoiling the statistical analysis that comes next.

9.1.4 Linear regression and ANOVA

The procedures for linear regression and ANOVA from the last chapter now find immediate application here. The main task is to pick out from the abundance of

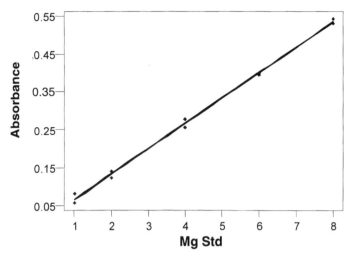

Figure 9.2 Dose–response line of the biuret assay for protein fitted by MINITAB linear regression through the tabulated points

possible outputs, the particular parts that are actually needed. Following §8.4.2, go to **Stat** > **Regression** > **Fitted Line Plot** and, when the dialog box opens, fill the spaces as:

---

1) *Response [Y]:* ['Abs Std'];
2) *Predictor [X]:* ['mg Std'] ;
3) *Type of Regression Model*
   [●] *Linear*;
4) Open the [Options] and [Storage] sub-windows and check that no additional output is requested meanwhile; then return to the main box;
5) [OK].

This gives Figure 9.2 as the linear regression dose–response line. It fits closely to the experimental points but note that it does not go down to the origin, of zero dose and absorbancy. This is because MINITAB does not plot graphs beyond the range of the data submitted, which is why the zero values had to be put in temporarily for the point plot in Figure 9.1.

Opening the Session window (Table 9.3) reveals the regression equation in **Section A** as:

$$Abs\ Std = -0.00168 + 0.0670\ mg\ Std \qquad \text{Eq. 9.1}$$

which can be generalized to:

$$y = -0.00168 + 0.0670\ x \qquad \text{Eq. 9.2}$$

- Note that this best-fitting line does not go exactly through the origin, but at a negative absorbance of $y = -0.00168$, when the dose $x = 0$. However, the t-test in

Table 9.3 Regression analysis, and analysis of variance, of the standard curve data in Cols 1 and 2 of Table 9.2

```
Section A
Regression
The regression equation is
Abs Std = - 0.00168 + 0.0670 mg Std


Section B
Predictor            Coef         StDev        T          P
Constant             -0.001677    0.006113     -0.27      0.791
mg Std               0.067018     0.001243     53.93      0.000


Section C
S = 0.01006     R-Sq = 99.7%     R-Sq(adj) = 99.7%


Section D
Analysis of Variance
Source               DF       SS          MS          F          P
Regression           1        0.29464     0.29464     2908.66    0.000
Residual Error       8        0.00081     0.00010
Total                9        0.29545
```

Section B of Table 9.3 shows that the departure from the origin is not significant. Thus the *'constant'* (−0.00168 in Eq. 9.2) has a *P*-value of 0.791 which is non-significant (i.e. not significantly different from zero). The other 'predictor', the *mg Std* is the 0.0670 of Eq. 9.2 and is highly significantly ($P = 0.000$) different from zero. It is a fundamental requirement of any assay that the dose–response line should, as here, have a significant slope.

- For later use, the *slope* of the dose–response line, $b = 0.0670$ needs to be highlighted, as does the equation itself.
- From **Section C**, is saved the *Error standard deviation*, $s = 0.01006$.

The ANOVA of the regression in Section D confirms that the regression is highly significant ($P = 0.000$) It also provides an estimate of the error standard deviation, as the square root of the *residual error MS* = 0.00010. The square root of this is 0.01, but already a more exact value has been obtained as $s = 0.01006$, above.

9.1.5 Exact estimates for the Unknowns

To obtain exact estimates of the protein content of the Unknowns, it is not possible to use the regression provisions on MINITAB, as in the last chapter for predicting students' fourth-year marks from their third-year marks. This is because *MINITAB only allows the prediction of y-values, and their confidence intervals, from chosen x-values*. In the present study, the opposite is wanted – the prediction of *x*-values (mg), and their confidence intervals from chosen *y*-values (absorbancies). Therefore the way forward is to use the regression equation (Eq. 9.1) rearranged as:

$$mg \ Std = (0.00168 + Abs \ Std)/0.0670 \qquad \text{Eq. 9.3}$$

and substituting into it the values from Table 9.2. Therefore in place of *Abs Std* will be entered the mean absorbancies of the Unknowns (C6, *Umean A*), and the output

(C7, *U mg*) will be in mg of protein in 1 ml of each Unknown. To do this, go to **Calc** > **Calculator** and, when the dialog box appears, enter (Figure 9.3):

1) *Store result in variable*: 'U mg';

2) *Expression*: (0.00168 +'Umean A')/0.0670;

3) Click on OK to close, and observe the output in C7 (Table 9.2) of the worksheet. These values are close to those already obtained from the graph and tabulated above in the text.

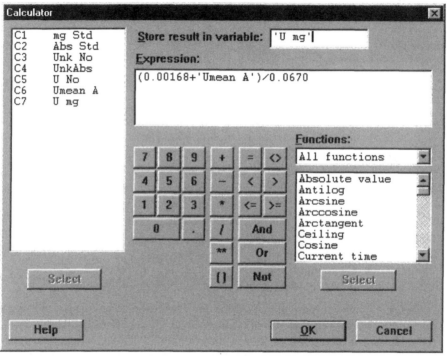

Figure 9.3 MINITAB dialog box for obtaining values of *dose* from values of *response* from the linear regression equation

9.1.6 Confidence intervals for the Unknowns

Except for very preliminary work, it is desirable that the output from an assay should include not only the *point estimates* as above, but also the 95% confidence intervals (CIs). With the student third- and fourth-year results, the CIs could be obtained directly from MINITAB, as described in §8.5.2. For the present assay, however, it is necessary to use MINITAB's *Calculator* function, for collecting basic statistics, some of which are already to hand.

The purpose of what follows is to calculate the 95% CIs from the equation

$$CI = \bar{x}_s + m \pm \frac{t.s.}{b} \sqrt{(1/N_s) + (1/N_u) + (m^2/SSx)} \qquad \text{Eq. 9.4}$$

Table 9.4 MINITAB worksheet (No. 2) for calculating the confidence intervals in the standard curve assay

C1	C2	C3	C4	C5	C6	C7	C8	C9	C10	C11	C12	C13
Xbar-S	Ybar-S	Umean A	b	m	t	s	Ns	Nu	RecipN	SSx	LCI	UCI
4.2	0.2798	0.046	0.067	−3.48955	2.306	0.01006	10	2	0.6	65.6	0.29448	1.12642
		0.280		0.00299							3.83946	4.56651
		0.501		3.28657							7.07619	7.89695

See text for explanation of column headings.

whose symbols are described below. First, it is convenient now to open another MINITAB worksheet (No. 2) within the present MINITAB project and label 13 columns as shown in Table 9.4. Most of these quantities will be obtained from the data already entered in worksheet No. 1 and, to avoid confusion, the columns in the two worksheets will be prefixed with W1 or W2. So that Col. 1 of worksheet No. 1 will be described as W1C1. It would, of course, be possible to do everything on the first worksheet, but this would involve the inconvenience of scrolling back and forth across the screen to access columns that are far apart. The entries on worksheet No. 2 are as follows:

- W2C1, *Xbar-S*: is the first term in Eq. 9.4 and is the mean of the 10 *mg Std* values in W1C1. It is obtained by entering this column into **Calc > Column Statistics** and, when the dialog box appears, responding:

> 1) Click on $\boxed{\bullet}$ *Mean*;
> 2) *Input variable:* $\boxed{\text{‘mg Std’}}$;
> 3) Leave unfilled *Store result in:* $\boxed{}$ *(Optional)*
> 4) Click on $\boxed{OK}$ to close the dialog box, and observe the output in the Session window as *Mean of mg Std* = 4.2000. Copy this into W2C1.

- W2C2, *Ybar-S*: is the mean of the *Abs Std* values in W1C2, and is obtained as for W2C1 but entering '*Ybar-S*' in step 2 above. From the output in the Session window, copy *Abs Std* = 0.27980 into W2C2, *Ybar-S*.
- W2C3, *Umean A*: is copied from W1C6.
- W2C4, *b*: is the slope of the regression line, already obtained as 0.0670.
- W2C5, *m*: is defined as:

$$m = (Umean\ A - Ybar\text{-}S)/b \qquad\qquad \text{Eq. 9.5}$$

and is obtained from **Calc > Calculator** and, when the dialog box appears entering:

> 1) *Store result in variable:* $\boxed{\text{‘m’}}$;
> 2) *Expression:* $\boxed{\text{(‘Umean A’-‘Ybar-S’)/0.0670}}$;
> 3) Click on $\boxed{OK}$ to close, and observe the output in W2C5 as a separate *m*-value for each unknown.

- W2C6, *t*: is the Student *t*-statistic for the degrees of freedom associated with *s*, the error standard deviation. Table 9.3 Section D shows that d.f. = 8 and, consulting a *t*-table gives 2.306 as *t* for 8 d.f. and a *P*-value of 0.05 (which is right for 95% confidence intervals).
- W2C7, *s*: is the error standard deviation = 0.01006 from Table 9.3, Section C, and was highlighted in the text above.
- W2C8, *Ns*: is the number of tests set up for the Standard preparation = 10.
- W2C9, *Nu*: is the number of tests set up for each Unknown preparation = 2.
- W2C10, *RecipN*: is the $(1/N_s) + (1/N_u)$ of Eq. 9.4 and equals 1/10 + 1/2 = 0.6.
- W2C11, *SSx* is the sums of squares of the *x*-values of the Standard in W1C1 and is obtained by going to **Calc > Calculator** and, when the dialog box appears, entering (Figure 9.4) the commands to execute:

$$SSx = \Sigma x^2 - (\Sigma x)^2/N \qquad \text{Eq. 9.6}$$

1) *Store result in variable*: $\boxed{SSx}$;

2) *Expression*: $\boxed{\text{SSQ('mg Std')-(SUM('mg Std')**2)/10}}$;

3) Click on $\boxed{OK}$ to close, and observe the output in W2C11 as 65.6.

- W2C12, *LCI*: is the destination column for the *lower* 95% CI of each unknown.
- W2C13, *UCI*: is the destination column for the *upper* 95% CI of each unknown.

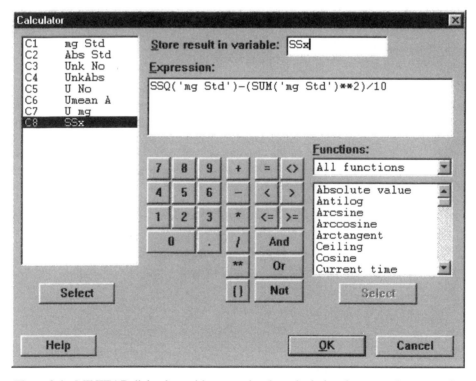

Figure 9.4   MINITAB dialog box with expression for calculating the sums of squares of *x*

Now all the quantities are available for substitution into Eq. 9.4, using the calculator function on MINITAB. The lower and upper confidence intervals are calculated separately with a change of sign at the '±' in the equation. To simplify the entry, the columns in W2 will be referred to by their column number, rather than by their abbreviated names as heretofore. Therefore go to **Calc** > **Calculator** and when the dialog box opens, enter the following:

---

1) *Store result in variable*: $\boxed{\text{C12}}$;
2) *Expression*: $\boxed{\text{C1+C5-(C6*C7/C4)*SQRT((C10)+((C5**2)/C11))}}$
3) Click on $\boxed{OK}$ to close, and observe that C12 now has the lower 95% CIs for the three Unknowns.

The upper 95% CIs are obtained similarly with *C13* as the destination column and the *Expression* altered by changing the – (minus) to a + after the first *C5*. This is done by editing that single character in the formula and without having to retype the whole expression. The final results for the three protein Unknowns are in Table 9.5.

Note (Col. 5) that the narrowest CIs are with Unknown B whose mean absorbancy was close to the middle of the dose–response curve. This had the effect of making the *m* in Eq. 9.4 close to zero and thereby having CIs whose width was minimal. Unknown A had very wide CIs, of ± 50% around the point estimate. This is partly because of the large value of *m* but more because the point estimate itself is so small that when CIs are expressed as a percentage, the range is large. Unknown C, with a similarly large *m*, had the same range of CIs as Unknown A, but the CIs as a percentage of the large point estimate were very narrow, only ± 5%.

9.1.7 Combining the standard deviations

It was assumed in the above calculations that the duplicate tests on the Unknowns would have the same *error standard deviation* (*s*) as the duplicate tests on the Standard. When explored by one-way ANOVA on Cols 3 and 4 of Worksheet 1, the error standard deviation (s_u) turns out to be 0.00736, somewhat lower than that of the Standard (s_s), which was 0.01006. For Eq. 9.4, it would be reasonable to use a value of *s* that combined the values from the standard, with 8 degrees of freedom (DF_s) and the unknowns, with 3 degrees of freedom (DF_u), since all the tests in the assay were presumably carried out with the same care. Therefore the two estimates of *s* can be combined to obtain a more representative value of the overall assay

Table 9.5 Final output of the calculations of point estimates and confidence intervals of the three protein Unknowns. Also shown are the ranges of the CIs, and the CIs as percentages around the point estimates

Unknown sample	Protein concentration (mg/ml) Point estimate	95%	CI	Range of CIs (UCI − LCI)	CIs as ±% of Point estimate
A	0.71	0.29	1.13	0.84	56
B	4.20	3.84	4.57	0.73	9
C	7.51	7.08	7.90	0.82	5

Table 9.6 Confidence intervals of Table 9.5 recalculated with an adjusted standard deviation, degrees of freedom and value of Student t

Unknown sample	Protein concentration (mg/ml) Point estimate	95%	CI	Range of CIs (UCI − LCI)	CIs as ±% of Point estimate
A	0.71	0.44	0.98	0.54	38
B	4.20	3.96	4.44	0.48	6
C	7.51	7.22	7.76	0.54	3.6

error, and with 11 degrees of freedom. The formula for combining the s-values is from the square root of the mean weighted error variances:

$$s = \sqrt{\{[DF_s.(s_s)^2 + DF_u.(s_u)^2]/(DF_s + DF_u)\}} \qquad \text{Eq. 9.7}$$
$$s = \sqrt{\{[8.(0.01006)^2 + 3.(0.00736)^2]/(8 + 3)\}}$$
$$= \sqrt{\{0.0009721/11\}}$$
$$= 0.0094009$$

This value for the combined *error standard deviations* is about 7% lower than that previously used, which will have a beneficial effect in narrowing the confidence intervals. A further consequence of combining the standard deviations is that the degrees of freedom are increased from 8 to 11, which reduces the Student-t from 2.306 to 2.201, which is also beneficial. Substituting these new values on the worksheet gives considerably narrower CIs, as detailed in Table 9.6, showing that the correction was worthwhile.

9.1.8 Exact 95% confidence intervals

The procedure for confidence intervals described above gives *approximate* 95% CIs. To obtain *exact* 95% CIs requires an extra factor g in Eq. 9.4. This additional factor is defined as:

$$g = (t.s/b)^2.(1/SSx) \qquad \text{Eq. 9.8}$$

and only becomes important if g is 0.05 or greater, otherwise it can be neglected. The values used to calculate the CIs in Table 9.6 were: $t = 2.201$; $s = 0.0094009$; $b = 0.067$ and $SSx = 65.6$. Substitution into Eq. 9.8 gives $g = 0.00145$, which is much less than the critical level at which g becomes important. Therefore with the biuret assay, g can be ignored. However, in bioassays, where the variability in responses may be much larger, g may be important and should therefore be considered. In the event that g has to be used, the equation for *exact* 95% CIs is:

$$CI = \bar{x}_s + \frac{m}{(1-g)} \pm \frac{t.s.}{b(1-g)} = \sqrt{(1-g).\{1/N_s) + (1/N_u)\} + (m^2/SSx)} \quad \text{Eq. 9.9}$$

This can be used in MINITAB's *Calculator* function, with the insertion of $(1 - g)$ at three places in the previous equation for approximate 95% CIs.

9.2 FOUR-POINT PARALLEL-LINE ASSAY

9.2.1 Preconditions and preliminaries

The simplest form of parallel-line assay is the four-point design, in which the responses given by two concentrations of a bioactive substance under investigation (the *unknown*) are compared with those of two concentrations of a *standard* preparation of the same substance. All four concentrations are tested with an equal number of replicates, and a common additional feature is the use of a randomized block design. Penicillin is the bioactive substance in the example to be described, and the response to it is the inhibition of bacterial growth due to antibiotic action. The assay is set up with the test bacteria uniformly inoculated into nutrient agar in Petri dishes, and four wells are cut in the agar of each dish. These are filled with a uniform volume of each of the four penicillin concentrations under test and the plates are incubated for 24 hours. During this time two competing processes occur: diffusion of the antibiotic out of the wells, and growth or inhibition of the bacteria in the agar medium. The result next day is circular and transparent growth-free zones around each well in culture plates where the rest of the agar is opaque with bacterial growth. The diameter of growth inhibition, which is the response measured, is a function of antibiotic concentration. Technical details may be found in Wardlaw (1982).

A necessary precondition to setting up a four-point assay is to determine the nature of the dose–response relationship with the standard penicillin preparation, so that appropriate concentrations of the standard may be chosen for the subsequent four-point assay. Table 9.7 shows a MINITAB worksheet with (C1, *Conc*) containing a range of eight penicillin concentrations, from 0.125 to 16 units/ml, in uniform twofold steps. The responses, in the form of zone diameters of growth inhibition, are tabulated in the next column (C2, *ZD*).

To investigate the dose–response relationship, go to **Stat > Regression > Fitted Line Plot** and, when the dialog box opens, fill the spaces as:

1) *Response [Y]*: ‘ZD’;
2) *Predictor [X]*: ‘Conc’;
3) *Type of Regression Model*
   ⦿ *Linear*;
4) Open the Options and Storage sub-windows and check that no additional output is requested meanwhile; then return to the main window;
5) OK.

Table 9.7 MINITAB worksheet for penicillin dose–response data. Concentration (Conc) is units/ml and zone diameter of growth inhibition is in arbitrary units of length

C1 Conc	C2 ZD	C3 LogConc
0.125	71.0	−0.90309
0.250	79.0	−0.60206
0.500	88.0	−0.30103
1.000	99.0	0.00000
2.000	107.5	0.30103
4.000	115.0	0.60206
8.000	128.0	0.90309
16.000	131.0	1.20412

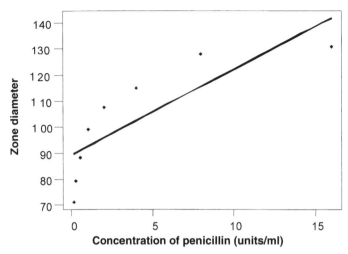

Figure 9.5   Untransformed plot of penicillin dose–response points, with linear regression line superimposed

This gives Figure 9.5 as the linear regression dose–response line. It is apparent that the actual relationship follows a hyperbolic curve, but that MINITAB has fitted a straight line through it. This illustrates the benefit of a graphical plot of the data at an early stage and before getting into numerical analysis. It is highly desirable, where possible, for a dose–response relationship to be a straight line and this can be done here by converting *Conc* to *Log Conc*. On MINITAB go to **Calc > Calculator** and, when the dialog box appears, enter:

1) *Store result in variable*: ⌐'LogConc'⌐;
2) *Expression*: ⌐LOGT('Conc')⌐;
3) Click on ⌐OK⌐ to close, and observe the output in Col. 3.

Repeating the above regression procedure, with *LogConc* instead of *Conc*, gives Figure 9.6, which shows a highly satisfactory linear dose–response line with closely fitting points.

$$\text{Zone diameter} = 97.8 + 29.8 \, \text{LogConc} \qquad \text{Eq. 9.10}$$

For the four-point assay itself, the two concentrations of standard penicillin selected were 1.25 and 5 units/ml, i.e. from the central region of Figure 9.6, and with a fourfold spacing of the concentrations.

9.2.2   MINITAB worksheet for four-point assay

The four-point assay itself was set up as a randomized block, with each of four culture plates being a 'block'. Each plate had one well each of the low and high concentrations (*doses*) of Standard and of Unknown. With the Unknown, the actual concentration of penicillin was to be determined in the assay, so the two 'doses' were of unknown concentration but the *low* dose was a fourfold dilution of the *high* dose, as for the Standard. The wells in each culture plate were charged

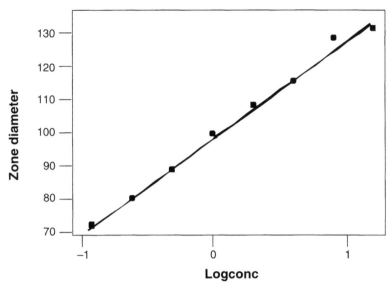

Figure 9.6 Linear regression of $\log_{10}$ concentration of penicillin against diameter of zones of growth inhibition

with the solutions in a pre-determined random sequence, to minimize possible bias due to time or positional effects.

Table 9.8 sets out the results of the four-point assay, with the four plates each delivering four readings of zone diameter. For MINITAB purposes, the data were entered on to a worksheet as in Table 9.9, which shows the first eight columns of a worksheet that is going to expand to 22 columns by the end of all the calculations. The first column

- **C1-T** has the extra suffix automatically added by MINITAB to indicate that it is a text column and not a data column. It shows the four culture plates labelled A–D.
- **Col. 2 *Prep*** lists the two preparations of penicillin by codes, where *1* = Standard and *2* = Unknown.
- **Col. 3 *Dose*** lists the concentrations of each preparation by codes, where *1* = lower dose and *2* = higher dose.
- **Col. 4 *ZD*** records the zone diameters of growth inhibition at the end of the assay.

Table 9.8 Sample results of a four-point parallel-line assay of penicillin

Plate	Zone diameter, arbitrary units (y)[a]			
	SL	SH	UL	UH
A	83	101	79	95
B	85	103	81	96
C	82	98	80	94
D	87	101	78	98

[a] Zone diameters were measured by placing each culture plate on an overhead projector and throwing an image, enlarged about 20 times, on to the laboratory wall. An ordinary ruler was then used to read the diameter of each zone in millimetres (Wardlaw, 1982).

Table 9.9 First section of MINITAB worksheet for the four-point parallel-line assay of penicillin

C1-T	C2	C3	C4	C5	C6	C7	C8
Plate	Prep	Dose	ZD	All P1	All P2	All D1	All D2
A	1	1	83	83	79	83	101
A	1	2	101	101	95	79	95
A	2	1	79	85	81	85	103
A	2	2	95	103	96	81	96
B	1	1	85	82	80	82	98
B	1	2	103	98	94	80	94
B	2	1	81	87	78	87	101
B	2	2	96	101	98	78	98
C	1	1	82				
C	1	2	98				
C	2	1	80				
C	2	2	94				
D	1	1	87				
D	1	2	101				
D	2	1	78				
D	2	2	98				

See text for explanation of column headings.

Thus each ZD value is identified by the coordinates of plate, preparation and dose of the first three columns. The results at this stage are ready for graphical evaluation.

9.2.3 Graphical evaluation

For MINITAB to plot the dose–response results, it is useful to open temporarily a new worksheet within the project and to copy into it the C3 (*Dose*) and C4 (*ZD*) data from the main worksheet as the first entries (Table 9.10). This is to avoid having an extra column in the first worksheet that will not be used later and which might be confusing. Next, a column is required (C3 *LogDose*) which contains the $\log_{10}$ of the *relative* doses in *volumetric* units, since the *actual* units per ml in the Unknown at this stage unknown. Each high dose of both Standard and Unknown is given a relative concentration value of 1.0, and each low dose becomes 0.25, since this is the relative volume of high dose solution per ml of low dose solution. These values are then converted to $\log_{10}$ to give -0.6026 for the $\log_{10}(0.25)$ and 0.0000 as the $\log_{10}(1.0)$. These two entries are then duplicated down C3 by going to: **Calc > Make Patterned Data > Arbitrary Set of Numbers** and, when the dialog box opens, filling the spaces as:

1) *Store patterned data in:* $\boxed{\text{'LogDose'}}$;
2) *Arbitrary set of numbers:* $\boxed{-0.6026 \ \ 0.0}$;
3) *List each value:* $\boxed{1}$ times;
4) *List whole sequence:* $\boxed{8}$ times;
5) $\boxed{OK}$.

Table 9.10 MINITAB worksheet for plot of four-point assay results

C1 Dose	C2 ZD	C3 LogDose
1	83	-0.60206
2	101	0.00000
1	79	-0.60206
2	95	0.00000
1	85	-0.60206
2	103	0.00000
1	81	-0.60206
2	96	0.00000
1	82	-0.60206
2	98	0.00000
1	80	-0.60206
2	94	0.00000
1	87	-0.60206
2	101	0.00000
1	78	-0.60206
2	98	0.00000

The point plot is then carried out as in §1.7.1 with the *Jitter* option, to yield the graph in the top part of Figure 9.7, which was edited to give that in the lower section, which shows:

- The points of Standard and Unknown in solid and open circles respectively; this was achieved by *unlocking the display* and changing the points of the Unknown, one by one, by reference to their tabulated *y*-values;
- Parallel straight lines fitted manually through the points to give two dose–response lines, which were labelled;
- The horizontal distance between the two dose–response lines established by extrapolating upwards from 0.0 on the dose axis to intercept the Unknown line and then across to Standard and back down to the axis.

This enables the $\log_{10}$ of the relative potency ($LogR$) to be read directly as -0.18. The $\text{antilog}_{10}(-0.18) = 0.66$.

Relative potency (R) is defined as:

$$R = \frac{\text{Volume of Standard}}{\text{Volume of Unknown}}$$

Eq. 9.11

that gives the same biological effect.

The 'biological effect' in this case is the particular zone diameter (*ca.* 96) of growth inhibition that was chosen by the extrapolation and insertion procedure. Since the Standard was known to contain 5.0 units/ml, the estimated potency of the Unknown is $5.0 \times 0.66 = 3.35$ units/ml.

Thus the graphical procedure allows an estimate of the *potency* of the Unknown, but gives no information on the *validity* of the assay. Nor is information provided on the 95% confidence intervals of the estimate.

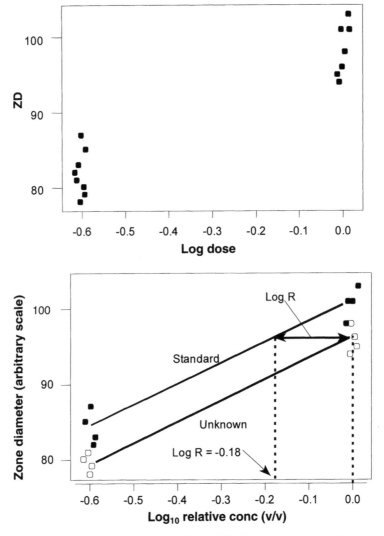

Figure 9.7 Four-point parallel-line assay of penicillin. Top, point plot of results; bottom, parallel dose–response lines fitted by eye to the same points and with the $\log_{10}$ relative potency (R) interpolated

9.2.4 Analysis of variance

Returning to the original worksheet, two separate ANOVAs are performed. The first is a *one-way ANOVA* to examine possible differences between plates in the assay. Go to **Stat > ANOVA > One-Way** and, when the dialog box opens, fill the spaces as:

1) *Response:* ZD ;
2) *Factor:* Plate ;
3) OK .

Table 9.11 MINITAB ANOVAS on results of penicillin assay: **A** to explore possible differences between plates; **B** to determine whether there are significant differences between the preparations, whether the slope is significant, and whether there is deviation from parallelism. Items for special attention are ⟨boxed⟩

Section A: Between-plate differences
One-way Analysis of Variance
Analysis of Variance for ZD

Source	DF	SS	MS	F	P
Plate	3	20.2	6.7	0.07	0.976
Error	12	1188.8	99.1		
Total	15	1208.9			

```
                          Individual 95% CIs For Mean
                          Based on Pooled StDev
Level    N    Mean     St Dev    ----------+----------+----------+----------
A        4    89.50    10.25     (----------*----------)
B        4    91.25    10.08          (----------*----------)
C        4    88.50    8.85      (----------*----------)
D        4    91.00    10.55        (----------*----------)
                                 ----------+----------+----------+----------
Pooled St Dev = 9.95                      84.0       91.0       98.0
```

Section B: Preparations, slope and parallelism
Two-way Analysis of Variance

Analysis of Variance for ZD

Source	DF	SS	MS	F	P
Prep	1	95.06	95.06	27.65	0.000
Dose	1	1072.56	1072.56	312.02	0.000
Interaction	1	0.06	0.06	0.02	0.895
Error	12	41.25	3.44		
Total	15	1208.94			

The output, which appears in the Session window, is recorded in the top part of Table 9.11. The important piece of information (boxed) is the *P*-value for *Between-Plate* differences. With $P = 0.976$, there is no evidence of such differences. If *P had* been significant here (<0.05), the assay would not have been invalidated, but a technical problem in the reproducible processing of the plates in the laboratory would have been detected. Such between-plate differences can arise, for example, if a Petri dish is inadvertently placed close to the blower fan in the incubator, so that evaporation of water from the agar is increased, and the antibiotic solutions are drawn into the agar faster than in plates placed elsewhere.

The second ANOVA is to investigate the main features of the assay itself and to obtain the *error standard deviation*. With the worksheet of Table 9.9 in the MINITAB window, go to **Stat > ANOVA > Two-Way** and, when the dialog box opens, enter:

1) *Response:* ⟨ZD⟩;
2) *Row factor:* ⟨Prep⟩;
3) *Column factor:* ⟨Dose⟩;
4) ⟨OK⟩.

Again, the output is in the Session window and is recorded, with salient features boxed, in the lower part of Table 9.11. Points to note are:

- $P = 0.000$ for *Prep* means that there is a highly significant difference between the two preparations, the Standard and the Unknown; the null hypothesis for *Prep* states that 'the two preparations differ in their zone diameters by no more than can be attributed to random-sampling fluctuations'; the $P \ll 0.05$ means that this statement can be rejected firmly and the contrary affirmed;
- $P = 0.000$ for *Dose* means that there is a highly significant difference between the low doses and the high doses of the two preparations taken together; in other words, the assay has a *highly significant slope*; here the null hypothesis states that 'the zone diameters produced by all the low doses differ from those of the high doses by no more than can be attributed to random-sampling fluctuations'; the $P \ll 0.05$ means that this statement can be rejected firmly and the contrary affirmed; it is essential for an assay to have a significant slope, and if this criterion is not met, the calculations should not proceed further;
- $P = 0.885$ for *Interaction* means that there is no significant deviation from the parallelism of the two dose–response lines; this is essential for the validity of a parallel-line assay and if not met, stops the calculations from going any further;
- The *Error MS* (mean square) $= 3.44$ provides the figure for getting the error standard deviation, s, which is needed later in the calculations; here $\sqrt{3.44} = 1.854\,72$.

Thus the analysis of variance provides assurance of validity (significant slope, parallel lines) of the results, together with a value for s.

9.2.5 Multiple sorting of the results

The graphical evaluation has already provided an estimate that the relative potency, $R = 0.66$. This now has to be confirmed and refined by calculation. Returning to the MINITAB worksheet in Table 9.9, the ZD results in Col. 4 have to be sorted out in two ways, first according to *Prep* and then according to *Dose*. Thus on the worksheet there is a column C5 *All P1*, in which are listed the eight ZDs produced by the low and high doses of Preparation 1, the Standard. Similarly, C6 *All P2* has all the ZD values for the Unknown. Likewise C7 *All D1* has the eight ZDs from all the low doses of Standard and Unknown lumped together, and C8 *All D2* has the eight ZDs from all the high doses.

This sorting was carried out by: Go to **Manip** > **Stack/Unstack** > **Unstack a Block of Columns** and, when the dialog box opens, enter:

1) *Unstack the following columns:* $\boxed{\text{ZD}}$;
2) *Using subscripts in:* $\boxed{\text{Prep}}$;
3) *Store unstacked data in blocks (one per subscript value):*
   | 'All P1' | ;
   | 'All P2' | ;
4) $\boxed{OK}$.

Inspection of the newly filled C5 and C6 on the worksheet shows that these two columns now contain the ZD values duly sorted *by Prep*.

The same procedure is used to sort the ZDs according to *Dose*, the entries being:

---

1) *Unstack the following columns:* $\boxed{\text{ZD}}$;
2) *Using subscripts in:* $\boxed{\text{Dose}}$;
3) *Store unstacked data in blocks (one per subscript value):*
   | 'All D1' |;
   | 'All D2' |;
4) $\boxed{OK}$.

Having sorted the ZD values in these two ways, it is now necessary to add them up and put the results in the next adjacent columns on the worksheet. The additions are performed by going to **Calc > Column Statistics** and, when the dialog box opens, entering the requests as:

---

1) Click on $\boxed{\bullet}$ *Sum*;
2) *Input variable:* $\boxed{\text{'All P1'}}$;
3) *Store result in:* $\boxed{\text{'C9'}}$ *(Optional)*
4) $\boxed{OK}$.

This delivers the total of the Standard zone diameters = 740 to C9 *Sum P1*. Similarly the three other sums are delivered to C10 *Sum P2*, C11 *Sum D1* and C12 *Sum D2*. These have been entered on Table 9.12 which is a rightwards extension of Table 9.9. Other factors that can now usefully be inserted are:

- C13 *s*, the error standard deviation = 1.85472;
- C14 *N*, the number of results in the whole assay = 16;
- C15 *LogD*, the $\log_{10}$ of the ratio: high dose/low dose = $\log_{10}(4.0) = 0.6021$.
- C20 *t*, the tabulated Student $t = 2.179$ for the 12 degrees of freedom associated with *s* and for a $P = 0.05$.

This has now prepared the worksheet for the final round of calculations of *Slope*, *Relative potency* (R), *Standard error of the relative potency* (S_{logR}) and *95% Confidence intervals*. These calculations are made with four formulae as listed in the next section.

9.2.6 Slope, relative potency, S_{logR} and 95% confidence intervals

Slope (*b*): the assembled numerical values on the worksheet are first used to calculate the average slope (*b*) of the two dose–response lines, which can

Table 9.12 Continuation of MINITAB worksheet from Table 9.9

C9	C10	C11	C12	C13	C14	C15	C16	C17	C18	C19	C20	C21	C22
Sum P1	Sum P2	Sum D1	Sum D2	s	N	LogD	b	LogR	R	SlogR	t	LCI	UCI
740	701	655	786	1.85472	16	0.6021	27.1965	−0.179251	0.661834	0.0242447	2.179	0.586030	0.747443

Table 9.13 MINITAB worksheet for combining the results of five independent assays. Some of the numbers have been shortened from their delivery by MINITAB

C1	C2	C3	C4	C5	C6	C7	C8	C9	C10	C11	C12	C13
Assay No	SLogR	W	R	LogR	WLogR	Sum W	Sum WLogR	Rbar	SLogRbar	t	LCI	UCI
1	0.0294	1156.9	0.66	−0.1793	−207.44	2625.1	−535.78	0.6250	0.01952	2.00	0.5713	0.6838
2	0.0764	171.3	0.64	−0.1909	−32.70							
3	0.0660	229.6	0.74	−0.1286	−29.52							
4	0.0392	650.8	0.55	−0.2575	−167.57							
5	0.0490	416.5	0.58	−0.2366	−98.54							

legitimately be put together since ANOVA found no significant departure from parallelism.

$$b = \frac{Sum\ D2 - Sum\ D1}{0.5\ N.\ log_{10}D}$$

Eq. 9.12

The formula is applied by: **Calc** > **Calculator** and, when the dialog box appears entering:

---

1) *Store result in variable:* $\boxed{b}$;

2) *Expression:* $\boxed{(C12\text{-}C11)/(0.5*C14*C15)}$;

3) Click on $\boxed{OK}$ to close, and observe the output in C16 *b* as 27. 1965.

Relative Potency (*R*): for which an approximate value 0.66 has already been obtained graphically is calculated as $Log_{10}R$ from:

$$Log_{10}R = \frac{Sum\ P2 - Sum\ P1}{0.5\ N.b}$$

Eq. 9.13

The formula is applied by: **Calc** > **Calculator** and, when the dialog box appears entering:

---

1   *Store result in variable:* $\boxed{LogR}$;

2) *Expression:* $\boxed{(C10\text{-}C9)/(0.5*C14*C16)}$;

3) Click on $\boxed{OK}$ to close, and observe the output in C17 LogR as -0.179251. Taking the antilog of this gives *R* = 0.6618, which rounds to *R* = 0.66, the same as the graph value.

Standard error of the relative potency (S_{logR}): is obtained from:

$$S_{logR} = (2s/b)\sqrt{(1/N)}.[1 + (Log\ R/Log\ D)^2]$$

Eq. 9.14

The formula is applied by: **Calc** > **Calculator** and, when the dialog box appears entering:

---

1) *Store result in variable:* $\boxed{SlogR}$;

2) *Expression:* $\boxed{(2*(C13/C16)*SQRT((1/N)*(1+(C17/C15)**2)))}$;

3) Click on $\boxed{OK}$ to close, and observe the output in C19 SlogR as 0.0242447.

95% Confidence intervals (*LCI, UCI*): are calculated from:

$$LCI,\ UCI = Antilog\ (\ Log_{10}\ R \pm t.\ S_{logR})$$

Eq. 9.15

They are calculated individually. To get the *Lower CI*, go to **Calc** > **Calculator** and, when the dialog box appears enter:

---

1) *Store result in variable*: $\boxed{\text{LCI}}$;

2) *Expression*: $\boxed{\text{ANTILOG(C17–C20*C19)}}$;

3) Click on $\boxed{OK}$ to close, and observe the output in C21 LCI as 0.5860.

The *Upper CI* is obtained similarly by:

1) *Store result in variable*: $\boxed{\text{UCI}}$;

2) *Expression*: $\boxed{\text{ANTILOG(C17+C20*C19)}}$;

3) Click on $\boxed{OK}$ to close, and observe the output in C22 UCI as 0.747443.

The CIs are converted to units/ml by multiplying by five to give the final result as:

Potency of Unknown = 3.3 u/ml, with 95% CIs 2.9 and 3.7 u/ml.

9.2.7 Combining several estimates

If there are several independent assays of the potency of a sample, the values of R and CIs can be combined to give a single best estimate of R and with much narrower CIs than from any of the individual assays. This is done not by taking the arithmetic average of the R-values, but by a process of *weighting*, which gives greatest emphasis to the assays of greatest precision (smallest S_{logR}), and least emphasis to those with the largest S_{logR}, which have lowest precision. Weight (W) is defined as:

$$W = 1/(S_{logR})^2 \qquad \text{Eq. 9.16}$$

Table 9.13 is a MINITAB worksheet for combining the results of five independent assays of the penicillin Unknown that has been under consideration. In Col. 3 it will be seen that the estimates of R go from 0.55 up to 0.74, a 30% range, while in Col. 2 the S_{logR} values are from 0.029 to 0.076, a three-fold range. The *weights*, however, show a sevenfold range because of the squaring of the S_{logR} values. Briefly, the procedure consists in proceeding across the worksheet with the *Calculator* function:

C6 W.LogR: is produced by multiplying C3 and C5;

C7 Sum W: is the sum of the W values in C3;

C8 SumW.LogR: is the sum of the $W.LogR$ values in C6.

C9 Rbar: is the antilog of the weighted mean estimate of the relative potency (R) as defined in:

$$R = \text{Antilog} \, [\Sigma WLogR/\Sigma W] \qquad \text{Eq. 9.17}$$

The expression for this in the *Calculator* dialog box is ANTILOG(C8/C7), with the output delivered to C9.

C10 SLogRbar: is the mean value of $S_{Log\,R}$ and defined by:

$$\overline{S_{logR}} = \sqrt{(1/\Sigma W)} \qquad \text{Eq. 9.18}$$

C11 t: is the value of Student's t for $P = 0.05$ and the total degrees of freedom for error variation. Since each assay has 12 d.f. associated with error, the total d.f. = 60 and the tabulated value of t is 2.00.

C12 *LCI*: is the lower 95% confidence interval and is calculated, together with the upper interval from:

$$95\% \text{ CI} = \text{Antilog}(\overline{Log_{10}R} \pm t. \overline{S_{logR}}) \qquad \text{Eq. 9.19}$$

The expression in the *Calculator* dialog box is ANTILOG(LOGT(C9)-C11*C10).

C13 *UCI*: is the upper confidence interval and is obtained similarly but with a + after (C9) in the previous expression.

The final result is a combined estimate of relative potency and confidence intervals:

$$R = 0.625 \text{ with } 95\% \text{ CIs } 0.57 \text{ and } 0.68.$$

After multiplication by five to convert to penicillin units/ml, these become 3.125 (2.85, 3.4) units/ml as the estimated penicillin concentration of the Unknown.

9.3 QUANTAL-RESPONSE ASSAY

9.3.1 Scope and problems

The two examples considered so far in this chapter involved *measurements* as the responses in the assays. A different approach is needed if the responses are *proportions*, such as 1/10, 6/10 and 9/10 for the proportions of animals killed in a toxicity assay with three doses of a test preparation. A commonly required output from such a test is the LD_{50}, or *lethal dose for 50%* of the animals. However, the response instead of being death may be some other 'all-or-nothing' change, such as the germination proportion of groups of 50 seeds, e.g. in relation to time of storage as the independent variable. Also the desired endpoint may not be at the 50% interpolate but at 10% or 90%. The term ED_{50} (effective dose-50) is the interpolated dose that is effective in 50% of the subjects, e.g in protecting against some disease. So there are many possibilities with such data, but the underpinning criterion is that each observed unit can exist in only one of two states such as dead/alive or affected/unaffected. The term *quantal-response* is applied to such assays where the *stimulus* leads to an all-or-nothing outcome in the *test subjects*. A classic example of such assays is the mouse-convulsion test for insulin, where the animals experience a transient (and non-injurious) loss of consciousness due to hypoglycaemia that can unambiguously be observed.

Statistically, this is an area that has many practical applications, not just in biology and medicine, but also in manufacturing processes where products such as light bulbs may pass/fail in relation to environmental stresses. It can be mathematically quite complicated and requires sophisticated treatment. Thus it is common for one or both of the variables in the assay to require transformation, as was needed with the *dose* variable in the penicillin assay. A further problem is that the dose–response line may have standard deviations that change systematically down the length of the line, flaring out towards one or both ends – the phenomenon of *heteroscedasticity* (§6.7.2). In consequence, when fitting the best straight line through the points, a formula for *weighted* regression has to be applied. This involves an *iterative* process, where the parameters of the line are refined by successive rounds of calculation, until the best fit is obtained. Such calculations can be done on an ordinary pocket calculator by formula, but only very tediously.

MINITAB (Release 12) is well endowed with programs for analysis of quantal data, but in what follows here only a small sampling of the available menus is provided. As working example, a data set on the titration of the virulence of bacteria in mice is taken. But similar data could have emerged from studies of seed germination in relation to time, or temperature, or humidity of storage; or the proportion of light bulbs still lit after 1000 hours, in relation to the environmental temperature of their surroundings.

9.3.2 Data for LD$_{50}$ estimation and graphical plot

Table 9.14 records the results of a bacterial virulence titration (assay) in mice. Dilutions of the challenge suspension from 10^{-2} to 10^{-7}, in tenfold steps, were injected into the animals, in groups of 15, and the numbers of deaths out of the total in each group were recorded. The lower section of the table shows how the data were entered on a MINITAB worksheet. It is assumed in the calculations that follow that:

- The animals were randomized before placement in the different dose groups, and were subsequently kept under identical conditions, i.e. between-group bias was minimized;
- The stated doses of infectious agent were administered without volumetric error;
- The susceptibilities of the animals to the infectious agent followed a lognormal distribution, as commonly observed; this means that a plot of percent survival (y) against log dose (x) should follow a cumulative frequency distribution curve, i.e. a symmetrical S-shaped curve.

A graphical plot of the data (Figure 9.8) suggests that the last assumption may be reasonable and also that the LD$_{50}$ is at a $\log_{10}$ dilution of around –5.2, or $10^{-5.2}$. The antilog of –5.2 is 0.0000063, or a dilution of approx 1/160 000. As will be seen below, this is not far from the value produced by calculation. However, the drawbacks to leaving the evaluation at this stage are that 1) the estimated LD$_{50}$ has no confidence intervals attached; and 2) there has been no analysis of the goodness-of-fit of the experimental points to the assumed underlying lognormal dose–response relation-

Table 9.14 Results of the virulence titration of a bacterial suspension in mice. Top, conventional tabulation; bottom, entries on MINITAB worksheet

Dilution of challenge suspension			10^{-2}	10^{-3}	10^{-4}	10^{-5}	10^{-6}	10^{-7}
No. dead/no. challenged			15/15	14/15	13/15	9/15	1/15	0/15
C1	C2	C3						
Dead	Total	Log10Dil						
15	15	–2.0						
14	15	–3.0						
13	15	–4.0						
9	15	–5.0						
1	15	–6.0						
0	15	–7.0						

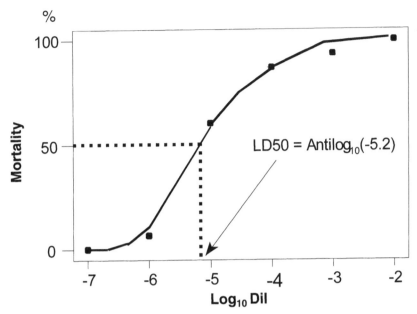

Figure 9.8 Sigmoid dose–response curve of bacterial virulence titration, with LD_{50} interpolated manually

ship and regression line. Therefore to explore these questions, further study is needed.

9.3.3 Probit transformation

The S-shaped dose–response curve can be converted into a straight line by the *probit* transformation, a standard procedure (but only one of several alternatives). This is desirable because it is more reliable to fit a straight line through experimental points, than to try to fit accurately an S-shaped curve.

The probit scale is in units of the standard deviation of the population in question, and with a central value of 5, as shown in Figure 9.9 (top). In drawing this curve, the S-shaped dose–response curve of the virulence titration was used to provide roughly estimated parameters ($\mu = -5.2$ and $\sigma = 0.6$) for the particular normal distribution plotted. For that reason, the *Log10Dil* scale of Figure 9.8 is included. In going from the normal distribution curve to the cumulative-frequency distribution curve, the dimensions of standard deviation and probit are preserved, as shown in Figure 9.9 (middle), which has *both* axes labelled with probits. But note that the vertical probit scale has unequal spacing of the tick marks, although they are symmetrically placed around the probit 5.

- A probit of 3 corresponds to the approx. 2.5% of the area (i.e. population) that has accumulated in the left 'tail' of the distribution;
- A probit of 4 corresponds to the approx. 16% of the left area of the distribution;
- A probit of 5 corresponds to one-half of the area (and of the population);
- A probit of 6 corresponds to about 84% of the area accumulated from the left;

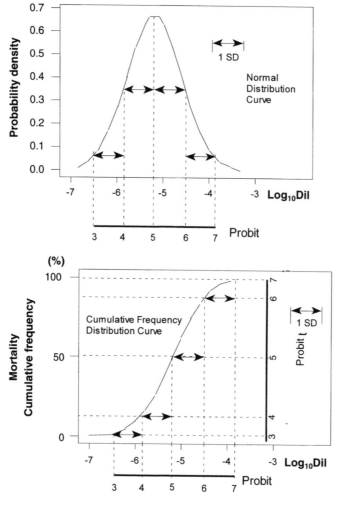

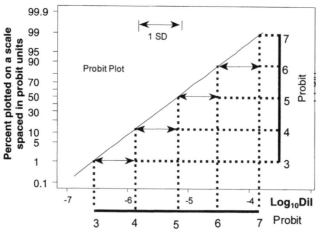

Legend overleaf

Figure 9.9 Diagrams to explain the application of probits to the bacterial virulence titration. Top, normal distribution curve with μ and σ (SD) roughly equal to those assumed in the probit transformation of the virulence titration data; middle, cumulative frequency distribution plot of the top diagram, showing that the chosen parameters fit closely the virulence dose–response curve in Figure 9.8; bottom, linearization of the virulence dose-response curve by transformation of percent mortality to probits

----

- A probit of 7 corresponds to about 97.5 of the area accumulated from the left and is near the top of the cumulative-frequency curve.
- There are no probit values for 0% or 100% because the normal distribution curve only reaches the axis at minus and plus infinity; thus for most purposes a probit scale from 2 to 8 is amply sufficient, as it covers the area of ±3 standard deviations on either side of the mean.

Finally, the bottom diagram for Figure 9.9 shows that the cumulative-frequency plot becomes a straight line when the vertical probit scale has its tick marks equally spaced. This has the effect of making the probability scale on the left ordinate have its tick marks spreading out above and below the 50% point. When the probit transformation is carried out on MINITAB, there is no presentation of an actual probit scale on the graph, but the ordinate percentage scale is as Figure 9.9 (bottom). A further point about the probits on MINITAB is that although they are in units of standard deviation, they are not centred on 5, but on zero.

9.3.4 Application of MINITAB

For MINITAB to analyse the virulence assay data by the probit method go to **Stat > Reliability/Survival > Probit Analysis** and, when the dialog box appears, enter the data as in Figure 9.10. This dialog box has an extensive supply of sub-dialog boxes, but here only the default options will be used:

----

1) ⦿ *Response in success/trial format*
   Number of successes: Dead ;
   Number of trials: Total ;
2) *Stress (stimulus)*: Log10Dil ;
3) Open the sub-dialog box for *Graph* and remove the checked request for:
   ☐ *Display confidence intervals*, leaving checked the single item
   ☑ Probability Plot
4) Click on OK to close the sub-dialog box and again on OK in the main box and observe the output as Figure 9.11.

The graph has had additional lines added to interpolate the LD_{50} dose as -4.9 on the $\log_{10}$ scale, corresponding to a dilution of about 1/80 000 or differing by a factor of 2 from the 1/160 000 read off the sigmoid dose–response curve. This difference highlights the desirability of getting the most objective estimates, and by calculation, rather than relying on the human eye to fit the most accurate intercept to the graph line. A further item of note is that although the sigmoid plot of the data

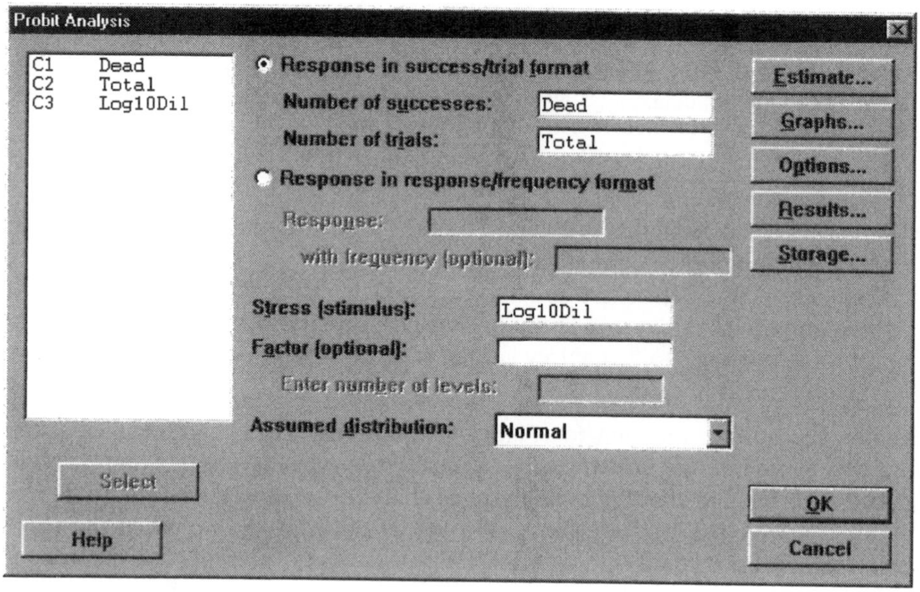

Figure 9.10   MINITAB dialog box for probit analysis of the bacterial virulence titration

(Figure 9.8) had six points corresponding to the six dilutions tested, the probit plot in Figure 9.11 only has four. This is because there is no probit value for 0 or 100% mortality. However, these values are not discarded from the calculation and do play a role in the iterative fitting of the probit straight line, albeit with very low *weight*.

Table 9.15   Edited printout of the MINITAB Session window from probit analysis of the virulence titration data

A: Regression Table

Variable	Coef	Error	Z	P
Constant	5.2852	0.9658	5.47	0.000
Log10Dil	1.0727	0.1949	5.51	0.000

Natural	Response	0.000

Log-Likelihood = −24.922

B: Goodness-of-Fit Tests

Method	Chi-Square	DF	P
Pearson	3.561	4	0.469
Deviance	3.178	4	0.529

C: Table of Percentiles (Log10 Dil)

Percent	Percentile	Error	95.0% Fiducial CI Lower	Upper
10	−6.1215	0.2970	−6.9559	−5.6566
50	−4.9268	0.1869	−5.3304	−4.5460
90	−3.7322	0.2755	−4.1702	−2.9701

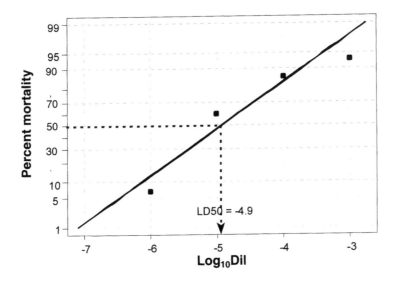

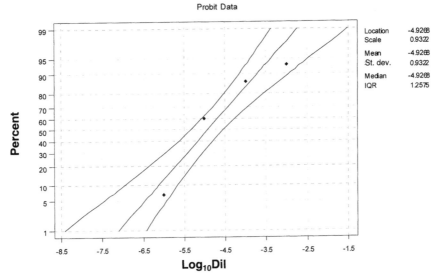

Figure 9.11 Probit versus $\log_{10}$ dilution of bacterial suspension. Top, linear regression line with LD_{50} interpolated by eye; bottom, the default linear regression plot, as delivered by MINITAB, with 95% confidence limits of the $\log_{10}$ dilution

Along with the dose–response graph of Figure 9.11, MINITAB also provides extensive output in the Session window, the salient points of which have been copied into Table 9.15, with some editing.

Section A is similar to that for the regression analyses from the measurement–response assays in §§9.1 and 9.2. Although the linear regression equation, $y = a +$

bx, is not explicitly provided in the MINITAB output, the item labelled 'Constant' is a and 'Log10Dil' is b. Moreover, since MINITAB does not (contrary to convention) have its probit scale with a central value of 5, this number has to be added to the *Constant* 5.2852 in the regression table to get a true probit. Therefore the regression equation is:

$$y = 10.2852 + 1.0727x \qquad \text{Eq. 9.20}$$

where y = Probit (of the mortality percentage) and $x = \log_{10}\text{Dil}$. To understand the intercept: the undiluted bacterial suspension would have a dilution of 1 and a $\log_{10}\text{Dil}$ of zero. This would give an intercept of $y = 10.2852$, which is over 5 standard deviations from the mean and therefore outside of the normal range of tabulation of the probit scale. It would correspond to >99.99% mortality.

The LD_{50} is at a value of x where $y = 5$, i.e.

$x = (5 - 10.2852)/1.0727$
$\quad = -4.9270$, which is close to the value given under 'Percentile' on the second last line of Table 9.15.

The Z-test values are statistics that test a and b for significance of difference from zero, the P being the probability of getting these Z-values if the null hypothesis (that they *are* zero) is correct. The P-values of <0.001 mean that both the intercept and slope terms are highly significantly different from zero

The *natural response* term is a potential correction factor if there had been any background death of animals unrelated to the bacterial challenge. The log-likelihood is similar to a chi-square value and is from the last iteration of the algorithm that fitted the regression.

Section B contains two goodness-of-fit tests that evaluate how well the regression fits the assumed 'model' of a probit/log dose relationship. The large values of P, which are far above 0.05 mean there is no problem of lack of fit.

Section C is only a small part of a large table that MINITAB provides in the Session window as a default option in the probit analysis. The items selected for inclusion are for estimating the LD_{10}, LD_{50} and LD_{90} values, which are likely to be of most interest.

- *Percent* refers to the percent mortality chosen for interpolation;
- *Percentile* is the $\log_{10}$ dilution corresponding to LD_{10}, LD_{50} and LD_{90};
- *Error* is the standard error of the estimates; notice how it is minimal for LD_{50} and flares out at LD_{10} and LD_{90}. The minimal error associated with the LD_{50} endpoint is the primary reason for choosing it;
- The two last entries are the lower and upper confidence intervals of the chosen LD parameter.

9.3.5 Final output

Converting the final output of Section C into reciprocal dilutions gives endpoints (with lower and upper confidence intervals), all to two significant figures, as:

LD_{10}	1/1 300 000	(1/9 000 000	1/450 000)
LD_{50}	1/84 000	(1/210 000	1/35 000)
LD_{90}	1/5 500	(1/1500	1/930)

This example is no more that a preliminary view of the possibilities for complication and refinement with quantal-response assays. Further information may be found in the MINITAB *User's Guide* and specialist sources, but the advice of an experienced statistician should also be sought.

9.4 SUMMARY

This final chapter of the book has provided samples of three common types of dose–response assays. The biuret test, which came first, was the simplest, as there was no transformation needed to obtain a linear dose–response line. Next in complexity was a penicillin assay in which the dose axis had to be converted to log dose in order to give a linear relationship. Finally came a quantal-response assay where linearization required the dose axis to be on a log scale and the response axis to be in probits. In each case the procedure was to produce a graphical plot to obtain a rough estimate of the required endpoint(s) and then to refine the estimate by calculation. Linear regression with analysis of variance was used to establish the validity of the assay, and estimates of endpoints, or relative potency, were provided with confidence intervals. If these assays were to be performed on a routine basis it would be highly desirable to have MINITAB 'Macros' written, so as to perform the calculations automatically, after entry of the data.

The final message to readers, who need to do these or other statistical calculations professionally and on a routine basis, is to advise establishing regular contact with an expert statistician. Hopefully what I have provided in these pages will allow a higher level discussion than might otherwise have been possible.

References

A'Brook, R. and Weyers, J. D. B. (1996). 'Teaching of statistics to UK undergraduate biology students in 1995'. *J. Biol. Educ.*, **30**, 281–288.

Banic, S. (1975). 'Prevention of rabies by vitamin C'. *Nature Lond.*, **258**, 153–154.

Cochran, W.G. (1950). 'Estimation of bacterial densities by means of the most probable number'. *Biometrics*, **6**, 105–116.

Finney, D.J., Latscha, R., Bennett, B.M. and Hsu, P. (1963). *Tables for Testing Significance in a 2×2 Table*. Cambridge University Press, Cambridge.

Herbert, D., Phipps, P.J. and Strange, R.E. (1971). 'Chemical analysis of microbial cells', pp. 209–344 in J.R. Norris and D.W. Ribbons (eds), *Methods in Microbiology*, Vol. 5B. Academic Press, London.

MINITAB Inc. (1997a). *User's Guide 1: Data, Graphics and Macros*. MINITAB Inc, Pennsylvania.

MINITAB Inc. (1997b). *User's Guide 2: Data Analysis and Quality Tools*. MINITAB Inc, Pennsylvania.

Morris, J.A. and Gardner, M.J. (1998). 'Calculating confidence intervals for relative risks (odds ratios) and standardized ratios and rates'. *Br. Med. J.*, **296**, 1313–1316.

Norman, R.L. and Kempe, L.L. (1960). 'Electronic computer solutions for the MPN equation used in the determination of bacterial populations'. *J. Biochem. Microbiol. Technol. Engng.*, **2**, 157–163.

Rowe, R., Todd, R. and Waide, J. (1977). 'Microtechnique for most-probable number analysis'. *Appl. Environm. Microbiol.*, **33**, 675–680.

Taylor, J. (1962). 'The estimation of bacteria by tenfold dilution series'. *J. Appl. Bacteriol.*, **25**, 54–61.

Taylor, L.R. (1961). 'Aggregation, variance and the mean'. *Nature Lond.*, **189**, 732–735.

Tuomilehto, J., Tuomilehto-Wolf, E., Virtala, E. and LaPorte, R. (1990). 'Coffee consumption as trigger for insulin dependent diabetes mellitus in childhood'. *Br. Med. J.* **300**, 642–643.

Wardlaw, A.C. (1982). 'Four-point parallel-line assay of penicillin', pp.370–379, in: S.B. Primrose and A.C. Wardlaw (eds), *Sourcebook of Experiments for the Teaching of Microbiology*. Academic Press, London and New York.

Weiner, A.S. (1943). *Blood Groups and Transfusion*, 3rd edn, p. 190. Charles C. Thomas, Illinois.

Additional Reading

Armitage, P. and Colton, T. (eds) (1998). *Encyclopedia of Biostatistics,* 6 Vols. John Wiley & Sons, New York.

Berry, D.A. (1996). *Statistics, a Bayesian Perspective.* Duxbury Press, Belmont.

Clarke, G.M. (1994). *Statistics and Experimental Design,* 3rd edn. Arnold, London.

Finney, D.J. (1971). *Statistical Method in Biological Assay.* Griffin, London.

Jarvis, B. (1989). *Statistical Analysis of the Microbiological Analysis of Foods.* Elsevier, Amsterdam.

McCloskey, M., Blythe, S. and Robertson, C. (1997). *Quercus: Statistics for Bioscientists.* Arnold, London, Sydney, Auckland.

Michelson, S. and Schofield, T. (1996). *The Biostatistics Cookbook.* Kluwer Academic Publishers, Dordrecht.

Moore, D.S. and McCabe, G.P. (1999). *Introduction to the Practice of Statistics,* 3rd edn. Freeman, New York.

Ryan, B.F. and Joiner, B.L. (1994). *Minitab Handbook.* Duxbury Press, Belmont.

Sokal, R.R. and Rohlf, F.J. (1995). *Biometry: the Principles and Application of Statistics in Biological Research*, 3rd edn. Freeman, New York.

Watt, T.A. (1997). *Introductory Statistics for Biology Students,* 2nd edn. Chapman & Hall, London.

Weiss, N.A. (1995). *Introductory Statistics,* 4th edn. Addison-Wesley, Massachusetts.

Appendices: Statistical Tables and Figures

Appendix A1 Most probable numbers from three successive tenfold dilutions, five tubes of each

Numbers of tubes positive out of 5, at 3 successive dilutions			MPN per inoculum of the first dilution
0	1	0	0.18
1	0	0	0.20
1	1	0	0.40
2	0	0	0.45
2	0	1	0.68
2	1	0	0.68
2	2	0	0.93
3	0	0	0.78
3	0	1	1.1
3	1	0	1.1
3	2	0	1.4
4	0	0	1.3
4	0	1	1.7
4	1	0	1.7
4	1	1	2.1
4	2	0	2.2
4	2	1	2.6
4	3	0	2.7
5	0	0	2.3
5	0	1	3.1
5	1	0	3.3
5	1	1	4.6
5	2	0	4.9
5	2	1	7.0
5	2	2	9.5
5	3	0	7.9
5	3	1	11.0
5	3	2	14.0
5	4	0	13.0
5	4	1	17.0
5	4	2	22.0
5	4	3	28.0
5	5	0	24.0
5	5	1	35.0
5	5	2	54.0
5	5	3	92.0
5	5	4	160.0

Adapted from Taylor, J. (1962). *Reproduced by permission of the Society for Applied Bacteriology.*

Appendix A2   Most probable numbers from three successive tenfold dilutions, eight tubes of each

Numbers of tubes positive out of 8, at 3 successive dilutions			MPN per inoculum of the first dilution
8	8	7	208
8	8	6	139
8	8	5	98.2
8	8	4	70.2
8	8	3	51.0
8	8	2	38.5
8	8	1	30.1
8	8	0	24.0
8	7	8	59.6
8	7	7	50.8
8	7	6	43.3
8	7	5	36.9
8	7	4	31.4
8	7	3	26.7
8	7	2	22.6
8	7	1	19.1
8	7	0	15.9
8	6	6	28.4
8	6	5	25.0
8	6	4	21.8
8	6	3	18.9
8	6	2	16.3
8	6	1	13.8
8	6	0	11.5
8	5	6	21.3
8	5	5	18.9
8	5	4	16.6
8	5	3	14.4
8	5	2	12.3
8	5	1	10.3
8	5	0	8.42
8	4	5	14.8
8	4	4	13.0
8	4	3	11.1
8	4	2	9.40
8	4	1	7.74
8	4	0	6.22
8	3	5	11.8
8	3	4	10.2
8	3	3	8.67
8	3	2	7.18
8	3	1	5.82
8	3	0	4.67

(continued)

Appendix A2 *continued:*

Numbers of tubes positive out of 8, at 3 successive dilutions			MPN per inoculum of the first dilution
8	2	4	8.07
8	2	3	6.72
8	2	2	5.50
8	2	1	4.45
8	2	0	3.62
8	1	3	5.22
8	1	2	4.27
8	1	1	3.50
8	1	0	2.87
8	0	2	3.38
8	0	1	2.80
8	0	0	2.31
7	7	1	5.47
7	7	0	4.84
7	6	2	5.30
7	6	1	4.71
7	6	0	4.15
7	5	2	4.58
7	5	1	4.04
7	5	0	3.55
7	4	3	4.46
7	4	2	3.95
7	4	1	3.47
7	4	0	3.04
7	3	3	3.86
7	3	2	3.40
7	3	1	2.98
7	3	0	2.59
7	2	3	3.33
7	2	2	2.92
7	2	1	2.55
7	2	0	2.20
7	1	3	2.87
7	1	2	2.51
7	1	1	2.17
7	1	0	1.86
7	0	2	2.14
7	0	1	1.83
7	0	0	1.55
6	6	1	3.08
6	6	0	2.77
6	5	1	2.73
6	5	0	2.44

(continued)

Appendix A2 *continued:*

Numbers of tubes positive out of 8, at 3 successive dilutions			MPN per inoculum of the first dilution
6	4	2	2.69
6	4	1	2.41
6	4	0	2.14
6	3	2	2.38
6	3	1	2.11
6	3	0	1.86
6	2	2	2.09
6	2	1	1.84
6	2	0	1.60
6	1	2	1.82
6	1	1	1.58
6	1	0	1.35
6	0	2	1.56
6	0	1	1.34
6	0	0	1.13
5	5	1	2.07
5	5	0	1.85
5	4	1	1.84
5	4	0	1.63
5	3	2	1.82
5	3	1	1.61
5	3	0	1.41
5	2	2	1.60
5	2	1	1.40
5	2	0	1.21
5	1	2	1.39
5	1	1	1.20
5	1	0	1.01
5	0	2	1.19
5	0	1	1.01
5	0	0	0.83
4	4	0	1.28
4	3	1	1.27
4	3	0	1.10
4	2	1	1.09
4	2	0	0.93
4	1	2	1.08
4	1	1	0.92
4	1	0	0.76
4	0	2	0.91
4	0	1	0.75
4	0	0	0.60
3	4	0	1.01

(continued)

Appendix A2 *continued:*

Numbers of tubes positive out of 8, at 3 successive dilutions			MPN per inoculum of the first dilution
3	3	1	1.00
3	3	0	0.85
3	2	1	0.85
3	2	0	0.70
3	1	2	0.84
3	1	1	0.70
3	1	0	0.56
3	0	2	0.69
3	0	1	0.55
3	0	0	0.41
2	4	0	0.79
2	3	1	0.79
2	3	0	0.66
2	2	1	0.65
2	2	0	0.52
2	1	1	0.52
2	1	0	0.39
2	0	2	0.51
2	0	1	0.38
2	0	0	0.26
1	3	0	0.49
1	2	1	0.49
1	2	0	0.36
1	1	1	0.36
1	1	0	0.24
1	0	2	0.36
1	0	1	0.24
1	0	0	0.12
0	2	0	0.23
0	1	1	0.23
0	1	0	0.11
0	0	1	0.11

From Norman, R. L. and Kempe, L. L. (1960), *J. Biochem. Microbiol. Technol. Engng.*, **2**, 157–163.
Reproduced by permission of John Wiley & Sons Inc.

Appendix A.3 Significance of differences between proportions. Diagrams determine the significance of differences between proportions from 3 × 3 to 10 × 10 comparisons. Black areas: $P \leqslant 1\%$; dotted areas: $5\% \geqslant P > 1\%$; plain areas: $P > 5\%$ (difference not significant)

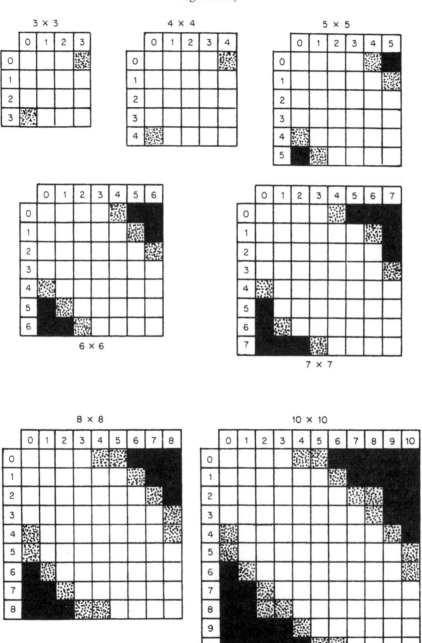

Appendix A.3 *continued:*

12 × 12

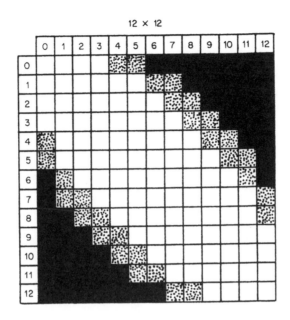

15 × 15

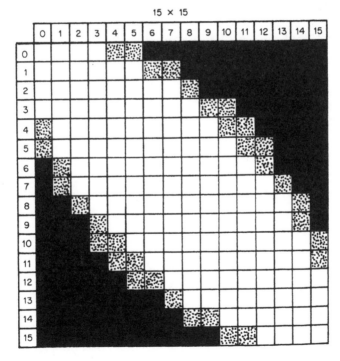

Index